Finite Elements

A Gentle Introduction

David Henwood
Department of Mathematics
University of Zimbabwe

Javier Bonet
Department of Civil Engineering
University of Wales, Swansea

MACMILLAN

First published 1996 by
MACMILLAN PRESS LTD
Houndmills, Basingstoke, Hampshire RG21 6XS
and London
Companies and representatives
throughout the world

ISBN 0-333-64626-6

A catalogue record for this book is available
from the British Library.

This book is printed on paper suitable
for recycling and made from fully
managed and sustained forest sources.

10   9   8   7   6   5   4   3   2   1
05   04   03   02   01   00   99   98   97   96

Typeset by EXPO Holdings

Printed in Hong Kong

Contents

Preface   xi

1   Finite elements introduced as bars forming a truss   1
   1.1   Introduction   1
   1.2   Definition of a truss, and the model problem   1
   1.3   Discretisation: nodes, elements and a numbering system   2
   1.4   Global and local coordinates   4
   1.5   Equations representing the mechanics of an element   5
   1.6   Changing from local to global variables   7
   1.7   Local and global variables   9
   1.8   Combining the equations   12
   1.9   Applying the boundary conditions   15
   1.10 Obtaining the solution   16
        General exercises for chapter 1   17

2   Some mathematical aspects   22
   2.1   Introduction   22
   2.2   The simple problem   22
   2.3   Mathematical formulations   23
   2.4   Exploring the idea of a functional   26
   2.5   The finite element method revealed   30
       2.5.1   Equations for the trial function   31
       2.5.2   Calculating the value of the functional   31
       2.5.3   Finding the optimum u_1 and u_2 32
   2.6   Approximate and exact solutions   33
        General exercises for chapter 2   34

3   Towards a systematic method   37
   3.1   Introduction   37
   3.2   More general boundary conditions   37
   3.3   Calculations element by element   39
   3.4   Calculations on a general element   40
       3.4.1   Forming the approximation for $u(x)$ 41
       3.4.2   Energy of the element   42
       3.4.3   Differentiation   42
   3.5   The finite element equations   43
   3.6   Three-element solution   44

Appendix 48
General exercises for chapter 3 49

4 The matrix approach **51**
4.1 Introduction 51
4.2 Casting the element equations in matrix form 51
4.3 The finite element equations 54
4.4 Forming the global matrix 55
4.5 Applying the boundary conditions 56
4.6 Displaying the solution 59
4.7 An axisymmetric problem in heat flow 59
    4.7.1 Calculations for a general element 61
    4.7.2 Forming the finite element equations 61
    4.7.3 Various boundary conditions 62
    General exercises for chapter 4 62

5 Two-dimensional heat flow **65**
5.1 Introduction 65
5.2 The model problem and an overview 65
5.3 Mathematical formulations 68
5.4 The simplified problem 70
5.5 Calculations for a general triangular element 71
    5.5.1 A general linear element formulation 71
    5.5.2 Computation for a particular 'general' element 74
    5.5.3 Forming the approximation for $u^e(\xi, \eta)$ 74
    5.5.4 Calculating V^e 76
5.6 The global finite element equations 77
5.7 The solution 77
    5.7.1 The element stiffness matrices 78
    5.7.2 Adding condensed **K** 80
5.8 Convection 82
5.9 Flux on the boundary 84
5.10 A point source of heat within the body 85
5.11 Summary 85
5.12 The model problem with numerical values 86
    General exercises for chapter 5 89

6 Variational form **93**
6.1 Introduction 93
6.2 Differential equation and variational form 93
6.3 Trial functions and residuals 94
6.4 The fundamental lemma of the variational calculus 96
6.5 Changing into a variational form 96
    6.5.1 The one-dimensional case 96
    6.5.2 Extension to two dimensions 97

6.6   Symmetry for trial and test functions   98
       6.6.1   The one-dimensional case with essential boundary
               conditions   100
       6.6.2   The two-dimensional case with essential boundary
               conditions   100
       6.6.3   Natural boundary conditions   101
6.7   Language of functionals   103
6.8   Connection between functionals *B*, *L* and *V*   104
       General exercises for chapter 6   108

7   The Galerkin approach   112
7.1   Introduction   112
7.2   A little about function spaces   112
7.3   Basis for element function spaces   113
7.4   Problem domain function spaces   115
       7.4.1   Forming a piecewise linear global approximation   116
       7.4.2   Forming a piecewise quadratic global approximation   118
       7.4.3   Basis functions for a two-dimensional piecewise
               linear global approximation   118
7.5   Using a general function space   119
7.6   The Galerkin method   122
7.7   The finite element method   122
7.8   Broadening the context of Poisson's equation   125
7.9   Summary of the mathematical setting of the finite element
       method   128
       General exercises for chapter 7   129

8   Element computation   133
8.1   Introduction   133
8.2   A simple illustrative example   133
       8.2.1   The illustrative problem   134
       8.2.2   What happens to the shape functions?   135
       8.2.3   What happens to differentiation with respect to x?   136
       8.2.4   What happens to 'dx'?   136
       8.2.5   The new integral and its evaluation   136
       8.2.6   The mapping formed by shape functions   138
       8.2.7   A revision of the key ideas   138
8.3   Gaussian quadrature   139
       8.3.1   On the interval $[-1,1]$ 139
       8.3.2   On the standard square   140
       8.3.3   On the standard triangle   141
8.4   Master elements and corresponding shape functions   142
       8.4.1   Three-noded master triangle   144
       8.4.2   Six-noded master triangle   145

		8.4.3	Four-noded square	147
		8.4.4	Eight-noded square master element	148
	8.5	Some element mappings		149
		8.5.1	Isosceles right-angled triangle mapped onto a general three-noded triangle	149
		8.5.2	Six-noded master element mapped onto a curved-sided triangle	150
		8.5.3	Eight-noded square mapped onto a curved-sided quadrilateral	151
		8.5.4	The general form of an element mapping	153
	8.6	Use of mappings in two-dimensional element computations		153
		8.6.1	A general look at mappings and their inverses	154
		8.6.2	The Jacobian and derivatives	154
		8.6.3	The effect of the mapping on areas	156
		8.6.4	What happens to shape functions?	156
		8.6.5	What happens to differentiation with respect to x and y?	157
		8.6.6	What happens to elements of area 'dx dy'?	158
		8.6.7	The new integral	158
	8.7	A look at the combined effect of individual mappings		159
	8.8	Mesh generation		160
		8.8.1	Mesh generation by mappings	161
			General exercises for chapter 8	162
9	**Elasticity**			**166**
	9.1	Introduction		166
	9.2	Background to elasticity		166
		9.2.1	Plane stress and plane strain	167
		9.2.2	Displacement and strains	167
		9.2.3	Stresses	169
		9.2.4	Equilibrium equations	170
		9.2.5	Relation between stress and strain	172
		9.2.6	Boundary conditions	172
		9.2.7	Summary	173
	9.3	Variational forms for elasticity		173
		9.3.1	Stating the problem	174
	9.4	A simple illustrative problem		177
	9.5	The general finite element approach and linear triangles		178
		9.5.1	The finite element approach	178
		9.5.2	The stiffness matrix for a general linear triangle	180
		9.5.3	Summary	186
	9.6	Working through the model problem with numerical values		186
			Appendix A	189
			Appendix B	190
			General exercises for chapter 9	191

Answers to exercises   196

Index   203

Preface

The method called 'finite elements' has been developed over the last thirty years into a popular technique for solving a number of significant problems in engineering and the physical sciences. The design office of an engineering firm or consultancy, whether for civil, mechanical or electrical engineering, will most probably have available software which can predict the performance of a proposed design while it is still on the drawing board. Displacements and stresses, heat or groundwater flow, magnetic or acoustic fields ... all now can be predicted with an accuracy which is usually acceptable.

The areas to which the method can be applied are ever widening; topics which today are non-standard and the subject of research in universities and research laboratories will be developed into tomorrow's software. Besides being actually used (or these days possibly because of it) the method is a topic in university courses. In addition to the utilitarian aspect, it draws together physics, mathematics, computer displaying of information, and has an elegance in its structure which is intellectually satisfying.

Both authors are currently working away from their home environments in countries full of interest (as are most countries) and with some features of special significance. Zimbabwe attracts tourists to see the grandeur of Victoria Falls, to visit the ruins of Great Zimbabwe set beneath high guarding rocks, and to see the game in the national parks. Swansea is near the Gower peninsula, an area of outstanding natural beauty. Every country has these features of special importance which are shown to tourists to give a brief overall impression; but there is much more to a country than can be shown in this way, and those who live and work there develop their own areas of interest and enjoyment. If finite elements were a country, this book would be a tourist guide as it discusses only the main features of the method. Working through it is like being a tourist (though sadly not as much fun) because the emphasis is on the basic concepts; the subdivision into elements, the formulation of the element and global equations and the mathematical background of the method. The justification for adding one more to the long list of 'Introduction to Finite Elements'-type of book lies in that here not many details are visited, but the basic ideas are treated carefully and worked through with numerical examples as far as possible. To continue with the tourist analogy, this book is not so much a coach trip as a walking tour with 'exercises' on the way to break the strain of absorbing other people's ideas and to fill in some details.

If you want to research the method or look in detail at particular aspects, you will need to refer to one of the several full treatments such as:

O. C. Zienkiewicz and R. L. Taylor (1994), *The Finite Element Method, Volumes 1–2*, 4th edition, McGraw-Hill.

E. B. Becker, G. F. Carey and J. T. Oden (1981), *Finite Elements, Volumes 1–6*, Prentice Hall.

T. J. R. Hughes (1987), *The Finite Element Method*, Prentice Hall.

The book grew out of lectures given at the (now) University of Brighton for students of engineering, computer science, and combined science in which mathematics is a main component, and it is hoped that it may prove useful for similar courses. Chapter 1 stands rather on its own; it may provide a useful introduction for engineering students who have met frameworks elsewhere, or it may be omitted. The other chapters form a dependent sequence with the last, on the application to elasticity, containing the most involved mathematics.

It has been found that learning can be made more interesting and students realise the usefulness of the method if *workshops* in a computer laboratory are given concurrently with the teaching of the theory. These sessions of say 2 hours can, using suitable software, illustrate and reinforce the ideas of the chapters. To encourage others to try this, a pamphlet containing a set of workshops to provide ideas, together with more detailed solutions to the exercises, is available on request from the publisher. If other lecturers have similar workshop ideas, they would be gratefully received and included; please e-mail henwood@maths.uz.zw or j.bonet@swansea.ac.uk.

Finally, we should thank those who have contributed to the development of this book: our students for teaching us how to teach, with no small generosity of spirit; and our colleagues for discussions, in particular John Barnes of the University of Brighton and Richard Wood from the University of Wales Swansea.

David Henwood, Department of Mathematics, University of Zimbabwe, Harare, Zimbabwe.

Javier Bonet, Department of Civil Engineering, University of Wales, Swansea.

1 Finite elements introduced as bars forming a truss

1.1 Introduction

Trusses are well known in engineering. In the process of solving for the displacements and stresses in bar members which make up these structures, some of the basic ideas of finite elements arise naturally. Because of this, the analysis of a simple truss will be used to introduce the method. The relevant ideas are:

- The division of a structure into elements and the devising of a numbering system for the elements and nodes.
- How a stiffness matrix relates displacements to the stresses (or strains) of an element.
- How the separate displacement/stress relationship for each element can be combined into a larger set of equations representing the whole structure.

A more mathematical approach, which introduces the additional concepts of function approximation and the minimising of functionals, will be given in later chapters.

1.2 Definition of a truss, and the model problem

A truss is a structure made up of straight members or bars connected at their ends by joints with smooth pins. As an example, consider the Howe roof truss which is shown in figure 1.1.

The division of a structure into **elements** happens very naturally when a truss is being analysed – each member is considered as a separate element. Generally, especially with solid structures of two or three dimensions, there is no natural shape or size for the elements, and the subdivision becomes something of an art

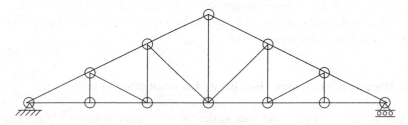

Figure 1.1   A Howe roof truss

1

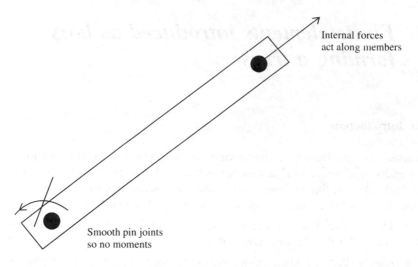

Internal forces
act along members

Smooth pin joints
so no moments

Figure 1.2   An idealised member or bar

(there are guidelines which it is wise to follow and these can, and have been, turned into computer software which will do the task for you).

The statement that a pin joint is 'smooth' implies that no friction acts between the pin and the member; consequently no moment is experienced by the member at the joint. All the internal forces act along the members and have no lateral components, see figure 1.2. Also, it is assumed that all the external loads acting on the structure, or the supports for the structure, are applied through the pin joints.

Figure 1.3 shows the simple model problem which will be analysed to provide this chapter's development of the finite element method. The structure is made up from four members of length l forming the square ABCD, with a fifth forming the diagonal BD. The members are connected by pin joints. The structure is fixed at A and has a roller joint at B; it is loaded by a force of magnitude P acting downwards at C at an angle of 45°. Let E be the modulus of elasticity (Young's modulus) and A be the common area of cross-section of the members.

Nodes are required by the method in order to define the geometry of the elements and hence the structure. A numbering system is needed to label these nodes and to label the elements themselves (and incidentally to be involved in the arranging of the linear equations to which the method gives rise). The pin joints will be used as nodes (figure 1.4).

1.3 Discretisation: nodes, elements and a numbering system

There are five elements and four nodes in the model problem. A numbering system has been chosen and is shown in figure 1.5. It is not unique; a different

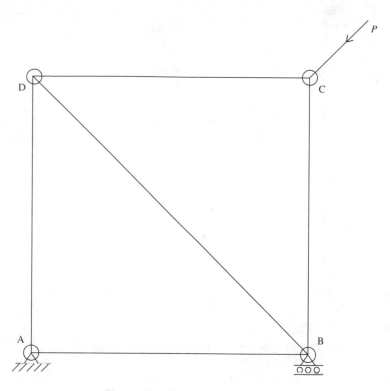

Figure 1.3   The model problem

choice would lead to a different ordering of the equations but of course it would still lead to the same solution. In a problem with many nodes, a good choice of numbering system can minimise the bandwidth of the finite element set of equations and hence reduce the computer storage requirements, but for small problems it makes little difference – just choose a sensible scheme. In this case element numbers are encased by parentheses ( ) and the node numbers are encircled.

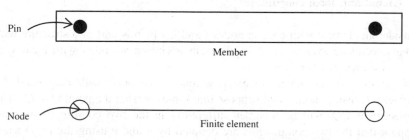

Figure 1.4   A bar and corresponding finite element

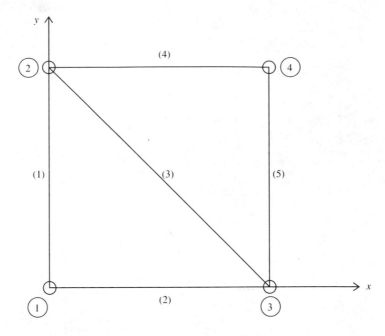

Figure 1.5   Node and element numbering

Coordinate axes are needed in order to define the geometry of the structure, the applied forces and the resulting displacements. So (x, y) axes have been chosen in the convenient position shown in figure 1.5, with the origin at the point A, node number 1. In section 1.4 new axes are going to be introduced which vary from element to element. They are aligned along and normal to each element and are chosen in order that the mechanical behaviour of each element can be expressed in a simple form. The variable axes are **local**, whereas the (x, y) set, fixed for the structure, are **global** axes.

1.4 Global and local coordinates

Consider an element with ends at nodes i and j, which is part of a structure with global coordinate axes (x, y). A local origin is chosen at node i with local axes (ξ, η) as shown in figure 1.6.

If the structure is subject to external loads each node will, in general, be displaced. Consider node j and suppose that it moves from the point Q to Q'. The displacement QQ' will be recorded differently in the two coordinate systems. Suppose that the two components are denoted by u and v using the (x, y) axes, and by u' and v' with the local (ξ, η) axes.

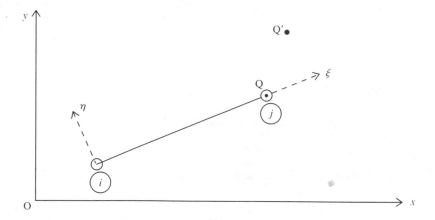

Figure 1.6   Global and local coordinates

An example will hopefully make this clear. Particular and arbitrary values for the displacement from Q to Q' in the two systems are shown in figure 1.7.

1.5 Equations representing the mechanics of an element

Suppose forces U'_i, V'_i act at node i with resulting displacements u'_i, v'_i, in the directions of the local axes (ξ, η), and similarly for node j, as shown in figure 1.8. Then from the equilibrium of the member

$$U'_i + U'_j = 0 \tag{1.1}$$

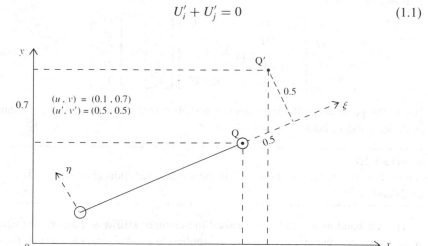

Figure 1.7   An example of displacements in local and global coordinates

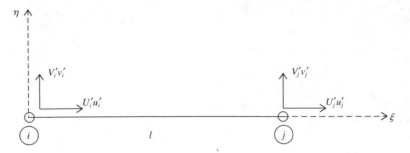

Figure 1.8   Forces and displacements in local coordinates

The change in the length of the member is $u'_j - u'_i$. If the displacements are small this implies a strain of $(u'_j - u'_i)/l$, while the stress is U'_j/A. Consequently, the theory of *strength of materials* gives

$$\frac{U'_j}{A} = E\frac{u'_j - u'_i}{l} \tag{1.2}$$

where l is the length of the element.

Also, it is assumed that there are no lateral forces, so

$$V'_i = V'_j = 0 \tag{1.3}$$

The finite element method is most naturally expressed using matrices, and adopting this approach the four equations contained in (1.1)–(1.3) are grouped into matrix form as

$$\frac{AE}{l}\begin{bmatrix} 1 & 0 & -1 & 0 \\ 0 & 0 & 0 & 0 \\ -1 & 0 & 1 & 0 \\ 0 & 0 & 0 & 0 \end{bmatrix}\begin{bmatrix} u'_i \\ v'_i \\ u'_j \\ v'_j \end{bmatrix} = \begin{bmatrix} U'_i \\ V'_i \\ U'_j \\ V'_j \end{bmatrix} \tag{1.4}$$

where the partition marks have been inserted to indicate a groupting associated with the i and j nodes.

Exercise 1.1
Verify that the matrix equation (1.4) is the same as the individual equations (1.1), (1.2) and (1.3).

The left-hand matrix in (1.4) is called the **element stiffness matrix**. The name arises from the relation between a force applied to a spring and the amount that it stretches,

stiffness × displacement = applied force

Note that the matrix has a particular form which is emphasised by the partitioning. If we write

$$\mathbf{M} = \begin{bmatrix} 1 & 0 \\ 0 & 0 \end{bmatrix},$$

then (1.4) becomes

$$\frac{AE}{l} \begin{bmatrix} \mathbf{M} & -\mathbf{M} \\ -\mathbf{M} & \mathbf{M} \end{bmatrix} \begin{bmatrix} u'_i \\ v'_i \\ u'_j \\ v'_j \end{bmatrix} = \begin{bmatrix} U'_i \\ V'_i \\ U'_j \\ V'_j \end{bmatrix}. \tag{1.5}$$

1.6 Changing from local to global variables

Displacements and the directions of forces are most simply expressed in terms of the local axes; however, there is a need to express the element equations with the displacements and forces in global, element-independent directions. This is necessary to ensure that when the element equations are combined the displacements and forces refer to the same directions. For example, node 4 in figure 1.5 belongs to the two elements 4 and 5 and consequently will have its displacement measured in different local coordinate systems, when each of the two elements is being considered. Consider element 2, which is shown in figure 1.9 with both local and global axes and variables, which happen in this case to be coincident.

Using the appropriate node and element numbers, equation (1.4) becomes, using local axes

$$\frac{AE}{l} \left[\begin{array}{cc|cc} 1 & 0 & -1 & 0 \\ 0 & 0 & 0 & 0 \\ \hline -1 & 0 & 1 & 0 \\ 0 & 0 & 0 & 0 \end{array} \right] \begin{bmatrix} u'_1 \\ v'_1 \\ u'_3 \\ v'_3 \end{bmatrix} = \begin{bmatrix} U'^2_1 \\ V'^2_1 \\ U'^2_3 \\ V'^2_3 \end{bmatrix} \tag{1.6}$$

A superscript has been added to the forces to indicate the corresponding element. Now since the local and global axes are in the same direction, $u'_1 = u_1$, $U'_1 = U_1$,

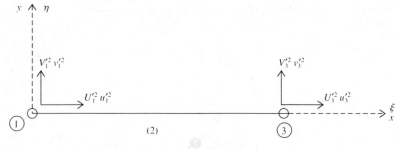

Figure 1.9   Element 2 with both local and global coordinates

etc., so (1.6) becomes

$$\frac{AE}{l} \left[\begin{array}{cc|cc} 1 & 0 & -1 & 0 \\ 0 & 0 & 0 & 0 \\ \hline -1 & 0 & 1 & 0 \\ 0 & 0 & 0 & 0 \end{array} \right] \left[\begin{array}{c} u_1 \\ v_1 \\ \hline u_3 \\ v_3 \end{array} \right] = \left[\begin{array}{c} U_1^2 \\ V_1^2 \\ \hline U_3^2 \\ V_3^2 \end{array} \right] \tag{1.7}$$

Exercise 1.2
Write down the corresponding element equations for element 4.

Element 1 is a bit more difficult.

The displacements and forces in terms of the local axes are shown in figure 1.10 for both nodes, 1 and 2, and in terms of the global axes for node 1 only.

The equations (1.1), (1.2) and (1.3) become

$$U_1'^1 + U_2'^1 = 0$$
$$\frac{U_2'^1}{A} = E\frac{u_2' - u_1'}{l} \tag{1.8}$$
$$V_1'^1 = V_2'^1 = 0$$

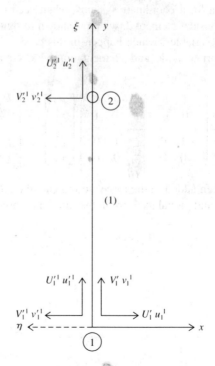

Figure 1.10   Element 1

Now, because of the relations between the directions of the local and global axes, for node 1

$$u_1' = v_1, \qquad v_1' = -u_1, \qquad U_1'^{1} = V_1^1, \qquad V_1'^{1} = -U_1^1.$$

Exercise 1.3
Write down the equations for node 2.

Rewriting the four equations (1.8) in terms of the global variables gives

$$V_1^1 + V_2^1 = 0, \qquad \frac{V_2^1}{A} = E\frac{v_2 - v_1}{l}, \qquad U_1^1 = U_2^1 = 0,$$

which may be combined with the results of exercise 1.3 into

$$\frac{AE}{l}\begin{bmatrix} 0 & 0 & 0 & 0 \\ 0 & 1 & 0 & -1 \\ 0 & 0 & 0 & 0 \\ 0 & -1 & 0 & 1 \end{bmatrix}\begin{bmatrix} u_1 \\ v_1 \\ u_2 \\ v_2 \end{bmatrix} = \begin{bmatrix} U_1^1 \\ V_1^1 \\ U_2^1 \\ V_2^1 \end{bmatrix} \tag{1.9}$$

Exercise 1.4
Write down the equations for element 5 corresponding to equation (1.9).

1.7 Local and global variables

A more serious difficulty occurs when element 3 is being considered, because the relationship between the directions of the local and global axes is not so simple. It is time to look at the problem more generally.

Consider a displacement QQ' represented by components in both the (ξ, η) and (x, y) directions and where the two sets of axes have a general orientation (see figure 1.11). Suppose the directions of the local axes are obtained from the global axes by rotation through an angle θ anticlockwise. Let the displacement vector have components u', v' in local coordinates and u, v using the (x, y) axes.

From simple trigonometry the following relationships may be obtained,

$$u = u' \cos\theta - v' \sin\theta \qquad v = u' \sin\theta + v' \cos\theta$$

i.e.

$$\begin{bmatrix} u \\ v \end{bmatrix} = \begin{bmatrix} \cos\theta & -\sin\theta \\ \sin\theta & \cos\theta \end{bmatrix}\begin{bmatrix} u' \\ v' \end{bmatrix}$$

or

$$\mathbf{u} = \mathbf{T}\mathbf{u}' \quad \text{where} \quad \mathbf{T} = \begin{bmatrix} \cos\theta & -\sin\theta \\ \sin\theta & \cos\theta \end{bmatrix} \quad \text{and} \quad \mathbf{u} = \begin{bmatrix} u & v \end{bmatrix}^{T}, \mathbf{u}' = \begin{bmatrix} u' & v' \end{bmatrix}^{T}.$$

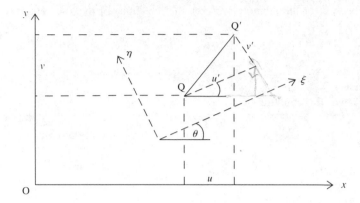

Figure 1.11   A displacement expressed in local and global coordinates

Exactly the same derivation can be given for the force vector $\mathbf{U} = [U, V]^T$, so that $\mathbf{U} = \mathbf{T}\mathbf{U}'$.

Thus we have the important result that

Multiplying by the matrix T transforms the components of a displacement (or force) from local into global coordinates.

Exercise 1.5

Show that $\mathbf{T}\mathbf{T}^T = \mathbf{T}^T\mathbf{T} = \mathbf{I}$, where $\mathbf{I}$ is the 2×2 unit matrix and $\mathbf{T}^T$ is the transpose of $\mathbf{T}$. As a consequence $\mathbf{T}^T = \mathbf{T}^{-1}$ and the matrix is said to be orthogonal.

Exercise 1.6

Deduce that

$$\mathbf{u}' = \mathbf{T}^T\mathbf{u} \tag{1.10}$$

Exercise 1.7

Show that the matrices $\mathbf{T}$ for the anticlockwise twist of $90°$ and the clockwise twist of $45°$ are

$$\begin{bmatrix} 0 & -1 \\ 1 & 0 \end{bmatrix} \quad \text{and} \quad \frac{1}{\sqrt{2}}\begin{bmatrix} 1 & 1 \\ -1 & 1 \end{bmatrix}, \quad \text{respectively.}$$

The change from local coordinates to global coordinates can be effected in the following way.

Recalling (1.5)

$$\frac{AE}{l}\begin{bmatrix} \mathbf{M} & -\mathbf{M} \\ -\mathbf{M} & \mathbf{M} \end{bmatrix}\begin{bmatrix} u_i' \\ v_i' \\ u_j' \\ v_j' \end{bmatrix} = \begin{bmatrix} U_i' \\ V_i' \\ U_j' \\ V_j' \end{bmatrix},$$

and using the relation between local and global coordinates (1.10), we have

$$\frac{AE}{l}\begin{bmatrix} \mathbf{M} & -\mathbf{M} \\ -\mathbf{M} & \mathbf{M} \end{bmatrix}\begin{bmatrix} \mathbf{T}^\mathrm{T}\begin{bmatrix} u_i \\ v_i \end{bmatrix} \\ \mathbf{T}^\mathrm{T}\begin{bmatrix} u_j \\ v_j \end{bmatrix} \end{bmatrix} = \begin{bmatrix} \mathbf{T}^\mathrm{T}\begin{bmatrix} U_i \\ V_i \end{bmatrix} \\ \mathbf{T}^\mathrm{T}\begin{bmatrix} U_j \\ V_j \end{bmatrix} \end{bmatrix}.$$

i.e.
$$\frac{AE}{l}\begin{bmatrix} \mathbf{M}\mathbf{T}^\mathrm{T} & -\mathbf{M}\mathbf{T}^\mathrm{T} \\ -\mathbf{M}\mathbf{T}^\mathrm{T} & \mathbf{M}\mathbf{T}^\mathrm{T} \end{bmatrix}\begin{bmatrix} u_i \\ v_i \\ u_j \\ v_j \end{bmatrix} = \begin{bmatrix} \mathbf{T}^\mathrm{T}\begin{bmatrix} U_i \\ V_i \end{bmatrix} \\ \mathbf{T}^\mathrm{T}\begin{bmatrix} U_j \\ V_j \end{bmatrix} \end{bmatrix}.$$

To simplify the right-hand side, multiply the first two and second two equations of the set by $\mathbf{T}$ and note that $\mathbf{T}\mathbf{T}^\mathrm{T} = \mathbf{I}$, giving finally

$$\frac{AE}{l}\begin{bmatrix} \mathbf{T}\mathbf{M}\mathbf{T}^\mathrm{T} & -\mathbf{T}\mathbf{M}\mathbf{T}^\mathrm{T} \\ -\mathbf{T}\mathbf{M}\mathbf{T}^\mathrm{T} & \mathbf{T}\mathbf{M}\mathbf{T}^\mathrm{T} \end{bmatrix}\begin{bmatrix} u_i \\ v_i \\ u_j \\ v_j \end{bmatrix} = \begin{bmatrix} U_i \\ V_i \\ U_j \\ V_j \end{bmatrix} \tag{1.11}$$

The last manipulation that produced (1.11) may seem fairly obvious, but unless you are experienced in this type of matrix manipulation you are advised to do the next exercise.

Exercise 1.8
Make sure you follow the manipulation leading to (1.11) (possibly by separating the equation into two using partition lines).

Returning to the problem of finding the stiffness matrix for element 3.
   Since the twist is given by $\theta = -45°$, see figure 1.12,

$$\mathbf{T} = \frac{1}{\sqrt{2}}\begin{bmatrix} 1 & 1 \\ -1 & 1 \end{bmatrix},$$

so
$$\mathbf{T}\mathbf{M}\mathbf{T}^\mathrm{T} = \frac{1}{2}\begin{bmatrix} 1 & 1 \\ -1 & 1 \end{bmatrix}\begin{bmatrix} 1 & 0 \\ 0 & 0 \end{bmatrix}\begin{bmatrix} 1 & -1 \\ 1 & 1 \end{bmatrix}$$
$$= \frac{1}{2}\begin{bmatrix} 1 & 1 \\ -1 & 1 \end{bmatrix}\begin{bmatrix} 1 & -1 \\ 0 & 0 \end{bmatrix}$$
$$= \frac{1}{2}\begin{bmatrix} 1 & -1 \\ -1 & 1 \end{bmatrix}.$$

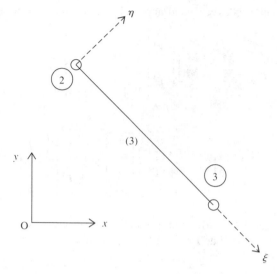

Figure 1.12   Element 3 with both local and global axes

Thus the element equation for 3 becomes, since the length is $\sqrt{2}l$,

$$
\frac{AE}{2\sqrt{2}l}
\begin{bmatrix}
1 & -1 & -1 & 1 \\
-1 & 1 & 1 & -1 \\
-1 & 1 & 1 & -1 \\
1 & -1 & -1 & 1
\end{bmatrix}
\begin{bmatrix}
u_2 \\
v_2 \\
u_3 \\
v_3
\end{bmatrix}
=
\begin{bmatrix}
U_2^3 \\
V_2^3 \\
U_3^3 \\
V_3^3
\end{bmatrix}.
\tag{1.12}
$$

Exercise 1.9
Show that $\mathbf{TMT}^{\mathrm{T}} = \begin{bmatrix} 0 & 0 \\ 0 & 1 \end{bmatrix}$ for a twist of $90°$, and compare with equation (1.9).

1.8 Combining the equations

The equations representing the behaviour of the individual elements have to be combined into a larger set which models the complete structure. Each pin (node) experiences forces from the members (elements) which it connects, and possibly from an external load. If there is no external force at a pin then the internal forces are in equilibrium; otherwise they balance the external load. The element equations are combined so that the right-hand sides represent the sum of the internal forces at a pin.

Collecting together the element equations written in terms of the global variables, we have

Element　　　　　　　　　　　　　　　　　　　　　Equation

1

$$\frac{AE}{l}\begin{bmatrix} 0 & 0 & 0 & 0 \\ 0 & 1 & 0 & -1 \\ 0 & 0 & 0 & 0 \\ 0 & -1 & 0 & 1 \end{bmatrix}\begin{bmatrix} u_1 \\ v_1 \\ u_2 \\ v_2 \end{bmatrix} = \begin{bmatrix} U_1^1 \\ V_1^1 \\ U_2^1 \\ V_2^1 \end{bmatrix}$$

2

$$\frac{AE}{l}\begin{bmatrix} 1 & 0 & -1 & 0 \\ 0 & 0 & 0 & 0 \\ -1 & 0 & 1 & 0 \\ 0 & 0 & 0 & 0 \end{bmatrix}\begin{bmatrix} u_1 \\ v_1 \\ u_3 \\ v_3 \end{bmatrix} = \begin{bmatrix} U_1^2 \\ V_1^2 \\ U_3^2 \\ V_3^2 \end{bmatrix}$$

3

$$\frac{AE}{2\sqrt{2}l}\begin{bmatrix} 1 & -1 & -1 & 1 \\ -1 & 1 & 1 & -1 \\ -1 & 1 & 1 & -1 \\ 1 & -1 & -1 & 1 \end{bmatrix}\begin{bmatrix} u_2 \\ v_2 \\ u_3 \\ v_3 \end{bmatrix} = \begin{bmatrix} U_2^3 \\ V_2^3 \\ U_3^3 \\ V_3^3 \end{bmatrix}$$

4

$$\frac{AE}{l}\begin{bmatrix} 1 & 0 & -1 & 0 \\ 0 & 0 & 0 & 0 \\ -1 & 0 & 1 & 0 \\ 0 & 0 & 0 & 0 \end{bmatrix}\begin{bmatrix} u_2 \\ v_2 \\ u_4 \\ v_4 \end{bmatrix} = \begin{bmatrix} U_2^4 \\ V_2^4 \\ U_4^4 \\ V_4^4 \end{bmatrix}$$

5

$$\frac{AE}{l}\begin{bmatrix} 0 & 0 & 0 & 0 \\ 0 & 1 & 0 & -1 \\ 0 & 0 & 0 & 0 \\ 0 & -1 & 0 & 1 \end{bmatrix}\begin{bmatrix} u_3 \\ v_3 \\ u_4 \\ v_4 \end{bmatrix} = \begin{bmatrix} U_3^5 \\ V_3^5 \\ U_4^5 \\ V_4^5 \end{bmatrix}$$

These equations are combined into a larger set of equations relating the set of all nodal displacements to the summed forces at the pins. For instance, the total horizontal force at node 1 has contributions from elements 1 and 2,

$$U_1^1 + U_1^2 = \frac{AE}{l}[0\ 0\ 0\ 0]\begin{bmatrix} u_1 \\ v_1 \\ u_2 \\ v_2 \end{bmatrix} + \frac{AE}{l}[1\ 0\ -1\ 0]\begin{bmatrix} u_1 \\ v_1 \\ u_3 \\ v_3 \end{bmatrix}.$$

These are brought together and related to all the displacements as,

$$\frac{AE}{l}[0+1\quad 0+0\quad 0\quad 0\quad -1\quad 0\quad 0\quad 0]\begin{bmatrix} u_1 \\ v_1 \\ u_2 \\ v_2 \\ u_3 \\ v_3 \\ u_4 \\ v_4 \end{bmatrix} = U_1^1 + U_1^2$$

Repeating this assembling process for all the nodes gives

$$
\frac{AE}{l}
\begin{bmatrix}
0+1 & 0+0 & 0 & 0 & -1 & 0 & 0 & 0 \\
0+0 & 1+0 & 0 & -1 & 0 & 0 & 0 & 0 \\
0 & 0 & \frac{1+0}{+\frac{1}{2\sqrt2}} & \frac{0+0}{-\frac{1}{2\sqrt2}} & -\frac{1}{2\sqrt2} & +\frac{1}{2\sqrt2} & -1 & 0 \\
0 & -1 & \frac{0+0}{-\frac{1}{2\sqrt2}} & \frac{1+0}{+\frac{1}{2\sqrt2}} & +\frac{1}{2\sqrt2} & -\frac{1}{2\sqrt2} & 0 & 0 \\
-1 & 0 & -\frac{1}{2\sqrt2} & +\frac{1}{2\sqrt2} & \frac{1+0}{+\frac{1}{2\sqrt2}} & \frac{0+0}{-\frac{1}{2\sqrt2}} & 0 & 0 \\
0 & 0 & +\frac{1}{2\sqrt2} & -\frac{1}{2\sqrt2} & \frac{0+0}{-\frac{1}{2\sqrt2}} & \frac{1+0}{+\frac{1}{2\sqrt2}} & 0 & -1 \\
0 & 0 & -1 & 0 & 0 & 0 & 1+0 & 0+0 \\
0 & 0 & 0 & 0 & 0 & -1 & 0+0 & 0+1
\end{bmatrix}
\begin{bmatrix}
u_1 \\ v_1 \\ u_2 \\ v_2 \\ u_3 \\ v_3 \\ u_4 \\ v_4
\end{bmatrix}
$$

$$
=
\begin{bmatrix}
U_1^1 + U_1^2 \\
V_1^1 + V_1^2 \\
U_2^1 + U_2^3 + U_2^4 \\
V_2^1 + V_2^3 + V_2^4 \\
U_3^2 + U_3^3 + U_3^5 \\
V_3^2 + V_3^3 + V_3^5 \\
U_4^4 + U_4^5 \\
V_4^4 + V_4^5
\end{bmatrix}
\qquad (1.13)
$$

These are the (global) finite element equations, and the matrix is the (global) stiffness matrix. They can be abbreviated as

$$\mathbf{Ku = f}$$

Consider the forces acting on a pin, say at node 1 (figure 1.13). $U_1^1, V_1^1, U_1^2, V_1^2$ are the components of the forces exerted by the pin on the member, so the forces *on the pin* are in the opposite direction.

The components of the force which fix the structure at node 1, X_1 and Y_1, must balance the internal forces, so that

$$X_1 = U_1^1 + U_1^2 \qquad Y_1 = V_1^1 + V_1^2$$

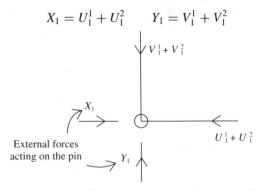

Figure 1.13   Forces acting on the pin at node 1

Similarly, at node 3, since there is no external force in the x-direction

$$0 = U_3^2 + U_3^3 + U_3^5 \qquad Y_3 = V_3^2 + V_3^3 + V_3^5$$

At node 2 there is no external force, so

$$0 = U_2^1 + U_2^3 + U_2^4 \qquad 0 = V_2^1 + V_2^3 + V_2^4$$

And at node 4,

$$P_x = U_4^4 + U_4^5 \qquad P_y = V_4^4 + V_4^5$$

where P_x and P_y are the components of the external applied load (as drawn in figure 1.3 both P_x and P_y are negative). Thus (1.13) becomes

$$\frac{AE}{l}\begin{bmatrix} 1 & 0 & 0 & 0 & -1 & 0 & 0 & 0 \\ 0 & 1 & 0 & -1 & 0 & 0 & 0 & 0 \\ 0 & 0 & 1+\frac{1}{2\sqrt{2}} & -\frac{1}{2\sqrt{2}} & -\frac{1}{2\sqrt{2}} & \frac{1}{2\sqrt{2}} & -1 & 0 \\ 0 & -1 & -\frac{1}{2\sqrt{2}} & 1+\frac{1}{2\sqrt{2}} & \frac{1}{2\sqrt{2}} & -\frac{1}{2\sqrt{2}} & 0 & 0 \\ -1 & 0 & -\frac{1}{2\sqrt{2}} & \frac{1}{2\sqrt{2}} & 1+\frac{1}{2\sqrt{2}} & -\frac{1}{2\sqrt{2}} & 0 & 0 \\ 0 & 0 & \frac{1}{2\sqrt{2}} & -\frac{1}{2\sqrt{2}} & -\frac{1}{2\sqrt{2}} & 1+\frac{1}{2\sqrt{2}} & 0 & -1 \\ 0 & 0 & -1 & 0 & 0 & 0 & 1 & 0 \\ 0 & 0 & 0 & 0 & 0 & -1 & 0 & 1 \end{bmatrix}\begin{bmatrix} u_1 \\ v_1 \\ u_2 \\ v_2 \\ u_3 \\ v_3 \\ u_4 \\ v_4 \end{bmatrix} = \begin{bmatrix} X_1 \\ Y_1 \\ 0 \\ 0 \\ 0 \\ Y_3 \\ P_x \\ P_y \end{bmatrix}$$

$$(1.14)$$

1.9 Applying the boundary conditions

The known applied force has already been included in (1.14). It remains to adjust the equations for the given restraints. Since node 1 is fixed, $u_1 = v_1 = 0$, and $v_3 = 0$ since the node 3 is restrained in the y-direction. Thus (1.14) represents eight equations with the eight unknowns $u_2, v_2, u_3, u_4, v_4, X_1, Y_1$ and Y_3 scattered in the left- and right-hand sides of the equations. In practice, it is usual to solve for the unknown displacements (ignoring the equations with the unknown reactions in the right-hand sides), and then later, if required, to use these equations to obtain the reactions at the restrained nodes.

Thus equations 1, 2 and 6 of (1.14) are initially omitted because they involve X_1, Y_1 and Y_3. Also, since the columns 1, 2 and 6 of the coefficient matrix multiply u_1, v_1 and v_3 which are zero, they are also dropped, leaving a square matrix. We have

$$\frac{AE}{l}\begin{bmatrix} 1+\frac{1}{2\sqrt{2}} & -\frac{1}{2\sqrt{2}} & -\frac{1}{2\sqrt{2}} & -1 & 0 \\ -\frac{1}{2\sqrt{2}} & 1+\frac{1}{2\sqrt{2}} & \frac{1}{2\sqrt{2}} & 0 & 0 \\ -\frac{1}{2\sqrt{2}} & \frac{1}{2\sqrt{2}} & 1+\frac{1}{2\sqrt{2}} & 0 & 0 \\ -1 & 0 & 0 & 1 & 0 \\ 0 & 0 & 0 & 0 & 1 \end{bmatrix}\begin{bmatrix} u_2 \\ v_2 \\ u_3 \\ u_4 \\ v_4 \end{bmatrix} = \begin{bmatrix} 0 \\ 0 \\ 0 \\ P_x \\ P_y \end{bmatrix}$$

$$(1.15)$$

1.10 Obtaining the solution

The set of equations (1.15) may now be solved to give,

$$u_2 = (2 + 2\sqrt{2})\alpha, \qquad v_2 = u_3 = \alpha, \qquad u_4 = (3 + 2\sqrt{2})\alpha, \qquad v_4 = \beta$$

where
$$\alpha = \frac{lP_x}{AE} \qquad \beta = \frac{lP_y}{AE}$$

To relate these to actual displacements we choose particular values for the material parameters, say $E = 2 \times 10^9$ N m^{-2}, $A = 0.0005$ m^2, and $l = 1$ m, so that $AE/l = 10^6$ N m.

Also, suppose that the applied load has components $P_x = P_y = -1000$ N, i.e. downwards and towards the fixed pin.

Then the displacements in mm are

$$u_2 = -4.83, \qquad v_2 = -1, \qquad u_3 = -1, \qquad u_4 = -5.83, \qquad v_4 = -1.$$

The forces at the restrained nodes may now be calculated from (1.14),

$$X_1 = -\frac{AE}{l} u_3 = 1000 \text{ N}$$

$$Y_1 = -\frac{AE}{l} v_2 = 1000 \text{ N}$$

$$Y_3 = -\frac{AE}{l} \left[\frac{1}{2\sqrt{2}} (u_2 - v_2 - u_3) - v_4 \right] = 0$$

The internal stresses in the members may be calculated from the element equations. For example, when considering element 3 and using (1.12), the forces at node 2 are given as

$$U_2^3 = +\frac{AE}{2\sqrt{2}l} (u_2 - v_2 - u_3 + v_3) = -1000 \text{ N}$$

$$V_2^3 = +\frac{AE}{2\sqrt{2}l} (-u_2 + v_2 + u_3 - v_3) = 1000 \text{ N}$$

In the direction of the local axes these become

$$\begin{bmatrix} U_2'^3 \\ V_2'^3 \end{bmatrix} = \mathbf{T}^{\mathrm{T}} \begin{bmatrix} -1000 \\ 1000 \end{bmatrix} = \frac{1}{\sqrt{2}} \begin{bmatrix} 1 & -1 \\ 1 & 1 \end{bmatrix} \begin{bmatrix} -1000 \\ 1000 \end{bmatrix}$$

$$= \frac{1}{\sqrt{2}} \begin{bmatrix} -2000 \\ 0 \end{bmatrix}$$

$$= -\sqrt{2} \begin{bmatrix} 1000 \\ 0 \end{bmatrix}$$

Exercise 1.10
Show that

$$\begin{bmatrix} U_3'^3 \\ V_3'^3 \end{bmatrix} = \begin{bmatrix} 1000\sqrt{2} \\ 0 \end{bmatrix}.$$

$1000\sqrt{2}$ $1000\sqrt{2}$

Figure 1.14   Forces acting on element 3.

So the member is in tension of magnitude $1000\sqrt{2}$ N (figure 1.14). (Compression is indicated by a negative local stress.)

Similarly, and more easily, it can be shown that all the other members are in compression of magnitude 1000 N.

Exercise 1.11
Verify the above statement.

General exercises for chapter 1

1. Discretise the K truss shown figure 1.15. Number the nodes and elements, and form a list of the nodes with their coordinates. Also, form a list of the elements, together with the two end nodes which define each element.

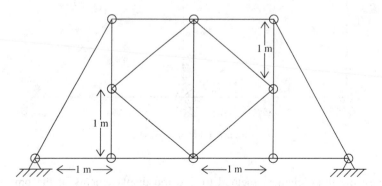

Figure 1.15   Exercise 1

2. In the simple Warren truss shown in figure 1.16, all the triangles are equilateral. Devise an element and node numbering scheme. Form the equations, in terms of global variables, for elements in the three essentially different positions (i.e. horizontal and slanting in two directions).

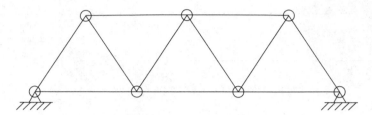

Figure 1.16   Exercise 2

3. Solve the chapter's model problem with a modified load, $P_x = 1000$ N, $P_y = 0$.

4. Solve the model problem with modified constraints $u_1 = v_1 = 0$, $u_2 = v_2 = 0$.

5. Consider the model problem extended by the addition of two members. Form the stiffness matrices for the two new elements, 6 and 7, shown in figure 1.17. What dimensions will the new global stiffness matrix have?

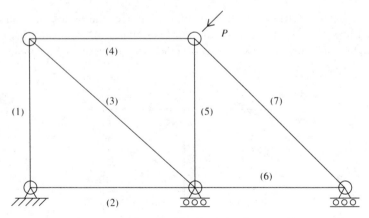

Figure 1.17   Exercise 5

6. Use the finite element method to find the displacements of the pins in the structure shown in figure 1.18. Each member is made of steel, $E = 2.0 \times 10^{11}$ N m^{-2}, and has cross-sectional area of 0.0005 m^2.

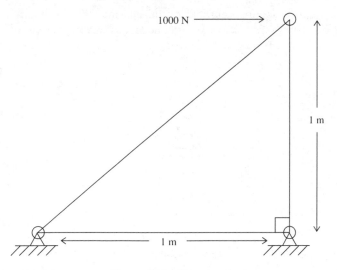

Figure 1.18   Exercise 6

7. The displacements for the truss shown in figure 1.19 are
$u_1 = 0.0,$ $v_1 = 0.0$ $u_2 = 17.48,$ $v_2 = -3.33$ mm.
$u_3 = -5.0,$ $v_3 = 0.0$ $u_4 = 22.48,$ $v_4 = -36.90$ mm.
Each member is made of steel, $E = 2.0 \times 10^{11}$ N m^{-2}, and has cross-sectional area of 0.0005 m^2. Calculate the axial forces in each member connected to node 2 and show that this joint is in equilibrium.

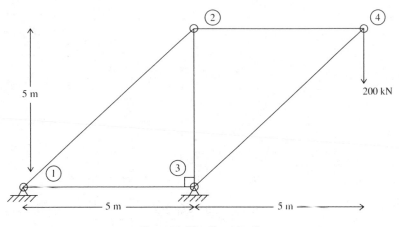

Figure 1.19   Exercise 7

8. For the structures shown in figure 1.20, calculate the nodal displacements and the axial force in each member, and verify that the axial forces produce a

system which is in equilibrium. Each member is made of steel, $E = 2.0 \times 10^{11}$ N m^{-2}, and has cross-sectional area of 0.0005 m^2.

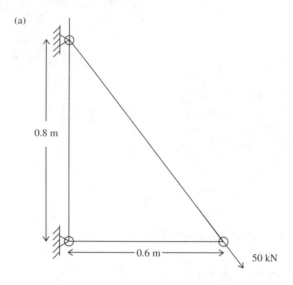

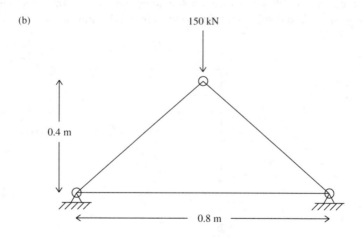

Figure 1.20   Exercise 8

9. Obtain the displacements and stresses in the three-member truss shown in figure 1.21. The member AB is 5 m long with area of cross-section

$A = 0.01 \, \text{m}^2$. BC is 8 m and $A = 0.02 \, \text{m}^2$, and for AC, $A = 0.01 \, \text{m}^2$. All the members are of steel with $E = 2.0 \times 10^{11} \, \text{N m}^{-2}$.

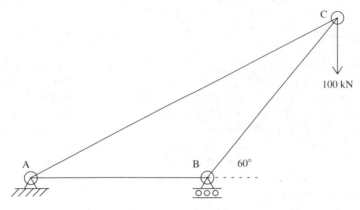

Figure 1.21   Exercise 9

2 Some mathematical aspects

2.1 Introduction

From the mathematical point of view the finite element method may be thought of as a technique for choosing the 'best' approximation to a solution from among a particular type or class of functions. These are called the **trial** functions. The criterion for the choice lies mainly with the physical laws which describe the problem, which are usually expressed as differential equations. The form of a differential equation is not immediately suitable for the finite element method; rather the problem has to be reformulated as a **variational principle**. In this chapter, where the setting is elasticity, the relevant variational principle is that of minimising the energy of the structure.

- The chapter illustrates the idea of a class of trial functions by the example of piecewise linear functions.

- The idea of a variational principle is considered, together with that of a **functional**, which in this context measures energy.

A simple problem is chosen to demonstrate the method and bring out these ideas.

2.2 The simple problem

A heavy uniform elastic string[1] BC of natural length l has modulus of elasticity E, weight per unit length w and area of cross-section A. Initially, the string BC lies horizontally (figure 2.1a), and then is moved to hang vertically from a point B' at the same level as B, with the lower end extended and fixed to a point C', so that $B'C' = l + b$. A general point on the string originally at P, where $BP = x$, is extended by an amount u to a point P'. Thus $B'P' = x + u$ in figure 2.1(b).

The extension of the point P to P', i.e. $B'P' - BP$, is caused by

(i) the weight of the string and

(ii) by the stretching resulting from its being attached at C'.

The extension will clearly vary according to the original position of the point P. There is no extension when P is at B; however, since $BC = l$ becomes $B'C' = l + b$, P at C is extended by b.

---

[1] An elastic grappling hook available in shops for securing luggage, etc. can demonstrate the situation. It is not completely satisfactory because, although it has weight, it is not sufficient to produce a noticeable extension.

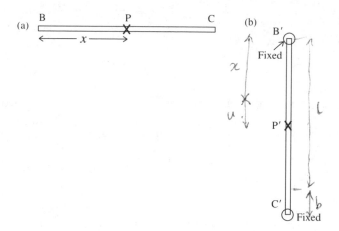

Figure 2.1   (a) Initial and (b) final positions of string

The problem is to find the extension $u(x)$ which, among other things, must satisfy the boundary conditions $u(0) = 0$ and $u(l) = b$ (figure 2.2).

2.3 Mathematical formulations

As mentioned in the introduction, the physical laws which the solution obeys usually involve rates of change, i.e. they are expressed as a differential equation. However, this form is not immediately suited to the finite element method, so a new (variational) form will be introduced. The differential equation is given for

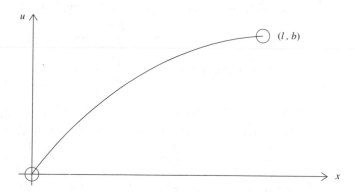

Figure 2.2   An extension with the two prescribed fixed points

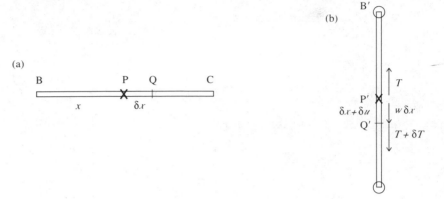

Figure 2.3   (a) Initial and (b) final positions of string

comparison. Both forms may be obtained by analysing the behaviour of a small section of the string, PQ, of length δx (figure 2.3a).

Suppose that as a result of the stretching of the string, P becomes P′ and Q becomes Q′ (figure 2.3b).

Let the extension of PQ be δu, so that $P'Q' = \delta x + \delta u$.

Let the tensions at P′ and Q′ be T and $T + \delta T$, respectively.

The weight of P′Q′ is $w\delta x$.

Initially, the traditional form of the physical laws is used to derive a differential equation.

1. **The extension, $u(x)$, as the solution of a differential equation.**

From the equilibrium of P′Q′ $T = T + \delta T + w\delta x$

so $$\frac{\delta T}{\delta x} = -w,$$

and letting $\delta x \to 0$ $$\frac{\mathrm{d}T}{\mathrm{d}x} = -w \qquad (2.1)$$

From stress $= E \times$ strain,

and since strain $\approx \delta u/\delta x$ for $\delta x, \delta u$ small,

$$\frac{T}{A} \approx E\frac{\delta u}{\delta x}$$

and letting $\delta x \to 0$ $T = AE\dfrac{\mathrm{d}u}{\mathrm{d}x} \qquad (2.2)$

Eliminating T between (2.1) and (2.2) gives the differential form

$$AE\frac{d^2u}{dx^2} = -w \quad \text{with boundary conditions,} \quad u(0) = 0, \quad u(l) = b \qquad (2.3)$$

Exercise 2.1

Show that the solution of (2.3) is

$$u(x) = -\frac{wx(x-l)}{2AE} + \frac{bx}{l}$$

Now for the variational form.

2. **The extension, $u(x)$, as the function which minimises the potential energy.**

The statement of the problem as a differential equation is about the behaviour of the problem variable 'at a point'. The rates of change and variables are related at a general point, and the equation must hold true wherever this general point is in the problem domain. In contrast, the variational form does not look at a particular point, but rather is a statement about the nature of the solution as a function defined over the problem domain. (Both approaches are of course eventually equivalent, as the solution at points builds up into a function defined over the domain and conversely the solution function can be evaluated at a general point in the domain.)

In order to state the problem in variational form we need to obtain a measure, for an arbitrary extension, of the energy contained in the stretched string $u(x)$. At this stage we need not specify anything about this arbitrary extension, or trial function, except that it should satisfy certain continuity conditions (required by the nature of the problem) and also the given boundary values. How the solution is identified from among the class of trial functions is the work of the variational principle. To return to measuring the energy of the string, this is found initially for the small section and then summed to give the energy of the whole string.

The potential energy of the element consists of two forms.

(a) Strain energy

Average of tension before and after stretching × extension.

$$\delta V_e = \frac{1}{2}\left(0 + AE\frac{\delta u}{\delta x}\right)\delta u = \frac{AE}{2}\left(\frac{\delta u}{\delta x}\right)^2\delta x \qquad (2.4)$$

(b) Gravitational energy

The weight of the element × distance below support

$$\delta V_g \approx -w\delta x(x + u) \qquad (2.5)$$

On adding (2.4) and (2.5), the total energy for P'Q' is

$$\delta V \approx \left[\frac{AE}{2} \left(\frac{\delta u}{\delta x} \right)^2 - w(x+u) \right] \delta x$$

Summing for the whole string and letting $\delta x \to 0$, the total potential energy stored in the string due to an arbitrary extension $u(x)$ is given by

$$V(u(x)) = \int_0^l \left[\frac{AE}{2} \left(\frac{du}{dx} \right)^2 - w(x+u) \right] dx \qquad (2.6)$$

We now have to invoke a law that will choose from the class of trial functions, $u(x)$, which one is the solution of the problem. The principle of minimum energy states that: **at equilibrium, the extension is such that the potential energy is a minimum.**

Thus the problem may be stated as

Find the extension $u(x)$ such that $u(0) = 0$ and $u(l) = b$
and which minimises $V(u(x))$

This is called a *variational* form of the problem, as the form or shape of $u(x)$ is considered to vary until it gives the minimum potential energy. $V(u(x))$ is called a *functional*; the thing which varies in it, $u(x)$, is itself a function.

The relation between a function and a functional may be expressed by:

'given a number, x, the *function u* determines a corresponding number $u(x)$'
'given a *function, u*, the *functional V* determines a corresponding number $V(u(x))$'

This will be illustrated in the next section.

Analytical techniques for finding such functions are part of the subject called the *Calculus of variations*, which is studied in numerous textbooks. The finite element method is a numerical procedure which approximates to the minimising function. It is widely applicable, whereas the analytic methods are able to produce a solution for only a very limited number of problems.

2.4 Exploring the idea of a functional

To avoid unnecessary manipulations, suppose that the following simplifying values are chosen to be set into (2.6); $AE = 2$, $w = l = b = 1$. Then the problem reduces to finding the function $u(x)$ such that $u(0) = 0$ and $u(1) = 1$, and which minimises

$$V(u(x)) = \int_0^1 \left[\left(\frac{du}{dx} \right)^2 - (x+u) \right] dx \qquad (2.7)$$

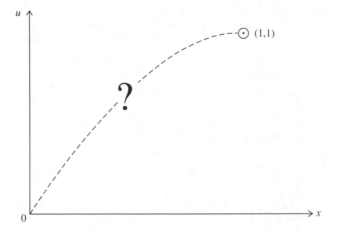

Figure 2.4   What is being sought

Viewed geometrically (figure 2.4), the problem is to find a curve passing through
the points (0,0) and (1,1), and which minimises (2.7). In order to develop the
idea of a functional and how functionals give rise to numbers, consider some
simple functions (curves) satisfying the boundary conditions, then calculate the
corresponding values of V.

(a) $V(u = x)$ (figure 2.5a)

$$= \int_0^1 [1^2 - (x + x)]dx$$

$$= 0$$

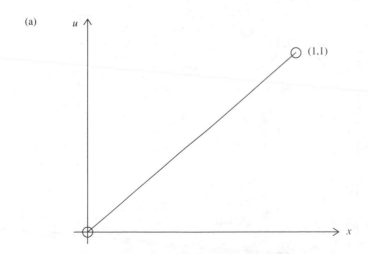

Figure 2.5   Simple curves (a)

(b)

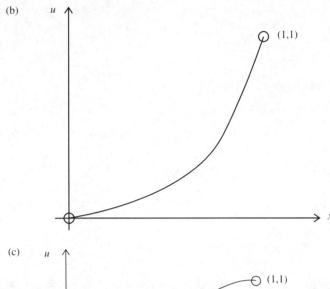

(c)

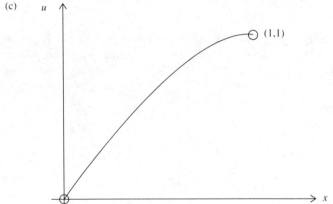

(d)

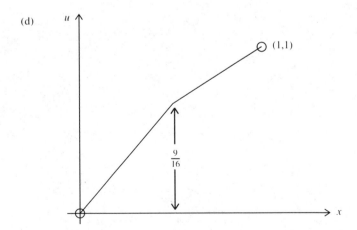

Figure 2.5   Simple curves (b,c,d)

(b) $V(u = x^2)$ (figure 2.5b)

$$= \int_0^1 [(2x)^2 - (x + x^2)]dx$$

$$= 0.5$$

(c) $V\left(u = \sin\frac{\pi}{2}x\right)$ (figure 2.5c)

$$\int_0^1 \left[\left(\frac{\pi}{2}\cos\frac{\pi}{2}x\right)^2 - x - \sin\frac{\pi}{2}x\right]dx$$

$$= 0.097$$

The next example may seem rather contrived, but it is in fact related to the finite element method. The 'curve' is constructed from two separate line segments joined together. The interval [0,1] is subdivided into two equal sections and the curve has a different equation on each.

$$u(x) = \begin{cases} \frac{9}{8}x & 0 \le x \le \frac{1}{2} \\ \frac{7}{8}x + \frac{1}{8} & \frac{1}{2} \le x \le 1 \end{cases}$$

(d) $V(u(x))$ (figure 2.5d)

$$= \int_0^{1/2} \left[\left(\frac{9}{8}\right)^2 - \frac{17x}{8}\right]dx + \int_{1/2}^1 \left[\left(\frac{7}{8}\right)^2 - \left(\frac{15x}{8} + \frac{1}{8}\right)\right]dx$$

$$= \frac{47}{128} - \frac{49}{128}$$

$$= -0.016$$

The particular choice for the value of u at $x = 1/2$, i.e. 9/16, calls for some explanation. By changing the mid-point value, a whole range (class) of two line-segment shapes (trial functions) can be obtained, varying from concave upwards, through straight, to concave downwards. The mid-point value 9/16 gives the minimum value of $V(u)$ that can be obtained by a member of the class. It will be shown later how the value can be calculated.

Exercise 2.2
Calculate $V(u)$ for $u(x) = \frac{x}{4}(5 - x)$. This is the solution found for the differential equation (2.3) and thus is the actual minimising function. The value of the functional is -0.021.

Exercise 2.3

Draw the graphs of $u(x)$ for (c), (d), and $u = \frac{x}{4}(5 - x)$, using the same axes for $0 \le x \le 1$. Consider the different shapes and the corresponding values of V.

2.5 The finite element method revealed

The finite element method is a systematic way of generating a particular class of trial functions and of obtaining the 'best' choice from the class. The function found will be the closest (in a sense) member of the class to the actual function which minimises V. Of course, if the exact solution belongs to the class, the method will find it.

To illustrate: consider the string subdivided into three sections or *elements* (figure 2.6). (The number of elements is not significant at this stage.) Suppose that the unknown values of u at the *nodes* P_1 and P_2 are denoted by u_1 and u_2. Each element and the internal nodes are numbered (the fixed end points, also nodes, have not been numbered in the example, but normally would be). For simplicity straight line segments or linear approximations over each element have been chosen (quadratics or higher order polynomials could be used). They combine together to form BP_1P_2C, a **piecewise** linear curve.

Note that this approximating curve is continuous, but that

- its derivative is not continuous, and
- it is defined by the two values u_1 and u_2, which have a physical meaning – they are the extensions at the internal nodes.

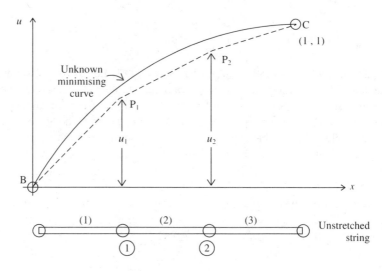

Figure 2.6   The string divided into three elements

We now seek the best trial function of the type of BP_1P_2C. To give structure to the manipulation, suppose that the work is broken down into stages (this is not the general finite element algorithm).

2.5.1 Equations for the trial function

The equations, expressed in terms of the (unknown) extension at the nodes, u_1 and u_2, are shown in (2.8). They are quoted rather than derived in order not to be side-tracked from the main theme; a modified approach which calculates the equation for a general element will be considered in chapter 3.

$$u(x) = \begin{cases} 3xu_1 & 0 \le x \le \frac{1}{3} \text{ element 1} \\ 3x(u_2 - u_1) + 2u_1 - u_2 & \frac{1}{3} \le x \le \frac{2}{3} \text{ element 2} \\ 3x(1 - u_2) - 2 + 3u_2 & \frac{2}{3} \le x \le 1 \text{ element 3} \end{cases} \quad (2.8)$$

Exercise 2.4

Verify that the piecewise linear curve defined by (2.8) is continuous and satisfies

$$u(0) = 0, \quad u(1/3) = u_1, \quad u(2/3) = u_2, \quad u(1) = 1.$$

2.5.2 Calculating the value of the functional

If the equations (2.8) are substituted into the functional V, an expression for V in terms u_1 and u_2 will result.

Consider the functional

$$V(u) = \int_0^1 \left[\left(\frac{du}{dx} \right)^2 - (x + u) \right] dx$$

Because the curve is defined by a different expression on each element, the integral is written in three parts:

$$V(u) = \int_0^{1/3} * \, dx + \int_{1/3}^{2/3} * \, dx + \int_{2/3}^1 * \, dx$$
$$= V^1 + V^2 + V^3, \quad \text{say} \quad (2.9)$$

which is the sum of energies stored in each element (where $*$ stands for the common integrand).

Using (2.8)

$$V^1 = \int_0^{1/3} [9u_1^2 - (x + 3u_1x)]dx = 3u_1^2 - \frac{1}{18} - \frac{u_1}{6}$$

$$V^2 = \int_{1/3}^{2/3} [(3u_2 - 3u_1)^2 - (x + 3x(u_2 - u_1) + 2u_1 - u_2)]dx$$

$$= 3(u_2 - u_1)^2 - \frac{1}{6} - \frac{u_1}{6} - \frac{u_2}{6}$$

Exercise 2.5

Show that

$$V^3 = 3(1 - u_2)^2 - \frac{4}{9} - \frac{u_2}{6}$$

Thus

$$V = V^1 + V^2 + V^3$$

$$= 3u_1^2 - \frac{1}{18} - \frac{u_1}{6} + 3(u_2 - u_1)^2 - \frac{1}{6} - \frac{u_1}{6} - \frac{u_2}{6} + 3(1 - u_2)^2 - \frac{4}{9} - \frac{u_2}{6} \quad (2.10)$$

2.5.3 *Finding the optimum* u_1 *and* u_2

It is important to note that by substituting in the equations for the trial function (2.8), the functional V has become a simple (quadratic) function in the variables u_1 and u_2. The well-established method for optimising using derivatives can now be employed. It is usual to find a stationary point and not to discriminate between a maximum or minimum because the physical situation indicates that only a minimum will exist.

The choice of u_1 and u_2 is made so that

$$\frac{\partial V}{\partial u_1} = \frac{\partial V}{\partial u_2} = 0$$

From (2.9)

$$\frac{\partial V}{\partial u_1} = 0 \Rightarrow \frac{\partial V^1}{\partial u_1} + \frac{\partial V^2}{\partial u_1} + \frac{\partial V^3}{\partial u_1} = 0$$

i.e.

$$6u_1 - \frac{1}{6} - 6(u_2 - u_1) - \frac{1}{6} + 0 = 0 \quad (2.11)$$

$$\frac{\partial V}{\partial u_2} = 0 \Rightarrow \frac{\partial V^1}{\partial u_2} + \frac{\partial V^2}{\partial u_2} + \frac{\partial V^3}{\partial u_2} = 0$$

i.e.

$$0 + 6(u_2 - u_1) - \frac{1}{6} - 6(1 - u_2) - \frac{1}{6} = 0 \quad (2.12)$$

Equations (2.11) and (2.12) when tidied up become

$$12u_1 - 6u_2 = \frac{1}{3}$$

$$-6u_1 + 12u_2 = \frac{19}{3}$$

which when solved give

$$u_1 = \frac{7}{18}, \qquad u_2 = \frac{13}{18}$$

Exercise 2.6
Evaluate the value of the minimising function $u(x) = \frac{x}{4}(5 - x)$ at $x = \frac{1}{3}$ and $x = \frac{2}{3}$ and compare with u_1 and u_2 obtained above.

Exercise 2.7
Obtain the corresponding value of V (using (2.10)).

2.6 Approximate and exact solutions

Clearly the finite element method does well (figure 2.7)! The good result in this case comes about because the exact solution happens to be quadratic. If it were a higher order polynomial, or a different function, perhaps the method would not be so impressive; but, generally, using more elements improves accuracy.

It is also instructive to compare the exact tension in the string with the tension obtained from the finite element solution.

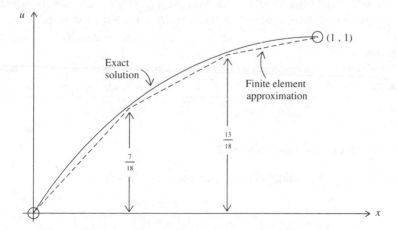

Figure 2.7 The exact and finite element solutions for u

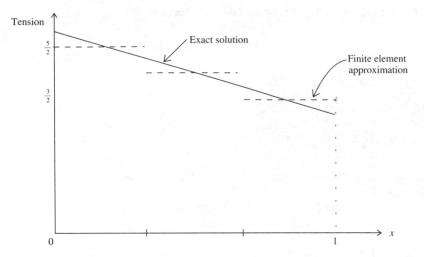

Figure 2.8 The exact and finite element solutions for the tension

From equation (2.2),

$$T = AE\frac{\mathrm{d}u}{\mathrm{d}x} = 2\frac{\mathrm{d}u}{\mathrm{d}x} \quad \text{since } AE = 2$$

Because the finite element solution is linear over each element, the resulting tension will be constant over each element, and is shown in figure 2.8. The finite element approximation to the tension, while it tries to follow the exact solution, is clearly limited by being piecewise constant (in fact the finite element solution averages the exact end values). This is a typical outcome of the finite element method; the problem variable, u in this case, is approximated closely, whereas the approximation deteriorates when the derivative is being considered.

Exercise 2.8
Rework the model problem using two elements, i.e. using only one unknown, u_1, which is u at $x = \frac{1}{2}$.

General exercises for chapter 2

1. A simple functional is one which measures length, i.e.

$$L\big(y = f(x)\big) = \int_a^b \sqrt{1 + y'^2}\,\mathrm{d}x,$$

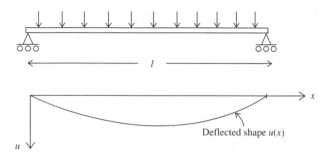

Figure 2.9   A simply supported uniform beam

where L is the length of the curve $y = f(x)$ between the two points $x = a$ and $x = b$. (The difficulty in asking many questions on L is that the integration is possible analytically for only a small number of functions.) Compare L between $x = 0$ and $x = 1$ for the two functions having common end points: (i) $f(x) = \cosh(x)$ and (ii) $f(x) = 1 + x(\cosh 1 - 1)$.

Of all possible functions which will give the minimum value?

2. A simply supported uniform beam is deflected by a uniformly distributed load, w per unit length (figure 2.9). The downward deflection $u(x)$ may be shown to satisfy the differential equation

$$EI\frac{d^2u}{dx^2} = \frac{w}{2}x(l - x) \qquad u(0) = u(l) = 0$$

where E is the modulus of elasticity and I the moment of inertia of a cross-section.

In the alternative variational formation, the deflection $u(x)$ satisfies $u(0) = u(l) = 0$ and minimises

$$V(u(x)) = \int_0^l \left[\frac{EI}{2}\left(\frac{du}{dx}\right)^2 + \frac{w}{2}x(l - x)u \right] dx$$

(a) Solve the differential equation to show that

$$EIu = \frac{w}{24}x(2lx^2 - x^3 - l^3)$$

(b) Use two elements as in exercise 2.8 to obtain a finite element solution.

(c) Compare the exact and finite element solution at $x = \frac{l}{4}$, $x = \frac{l}{2}$, and $x = \frac{3l}{4}$. Sketch the two solutions.

3. Consider a different functional, which arises in the design of pivots and bearings. The cross-section of a pivot is determined by a curve $y(x)$ that must pass through two fixed points $(0,1)$ and $(1,a)$ (figure 2.10). For the best

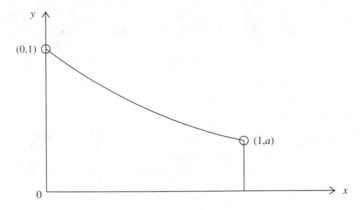

Figure 2.10   Cross-section of a pivot

performance the curve must minimise a functional V which is related to the friction experienced at the pivot. The function $b(x)$ is known:

$$V(y(x)) = \int_0^1 \left[y \left(\frac{dy}{dx} \right)^2 + b(x)y^2 \right] dx.$$

Choose $a = \frac{1}{2}$ for the exercise.

(a) Use two elements in the manner of exercise 2.8 to form a trial function satisfying the boundary conditions.

(b) Without attempting the detailed integration (which requires $b(x)$ anyway), show that the trial function turns V into a *cubic* in $y_1 = y(\frac{1}{2})$. The equation $dV/dy_1 = 0$ will then be a *quadratic* and hence will not have a unique solution.

(c) Usually, the problems to which the finite element method is applied give rise to quadratics for V, and hence linear equations to be solved for the nodal variables. Explain why the problem of this exercise would present difficulties.

3 Towards a systematic method

3.1 Introduction

The chapter makes a small extension to the work. The boundary conditions are generalised to include conditions on the derivative of the problem variable, and this results in further terms being added to the functional. The idea of a general element, to which each particular element may relate, is introduced. The ideas that the chapter emphasises are:

- That the computation can be based on an 'element by element' approach, which is used partly in this chapter and becomes more important later.

- By substituting the finite element trial function into the functional it is changed into an ordinary function of several variables, these being the values of the problem variable at the nodes.

- The computation may be carried out on a general element and then the results referred back to particular elements. In making the approximation for $u(x)$ on the general element, 'shape functions' are introduced to provide a neat and powerful means of interpolation.

3.2 More general boundary conditions

In the model problem the value of the unknown, the extension of the string, is given at both ends, i.e. $u(0) = 0$ and $u(l) = b$. The problem is now modified to allow for the possibility that the string is extended by means of a measured applied force at either, or both ends (figure 3.1). If the force is given, the extension at the corresponding end will be unknown initially, and will have to be calculated.

Various possibilities arise:

1. B', C' fixed, as in the model problem of chapter 2.
2. B' fixed and C' extended by known applied force T_C.
3. C' fixed and B' extended by a known applied force T_B.
4. Both B' and C' extended by known applied forces; though in this case equilibrium considerations require $T_B = T_C + wl$ where wl is the weight of the string.

When there are known applied forces, the functional representing the energy of the system has to be modified to include terms resulting from the work done by these forces. The functional (2.6) becomes

$$V(u(x)) = T_B u(0) + \int_0^l \left[\frac{AE}{2} \left(\frac{du}{dx} \right)^2 - w(x + u) \right] dx - T_C u(l) \qquad (3.1)$$

Figure 3.1   String and various boundary conditions

where $u(0)$ and $u(l)$ are the now unknown extensions at $x = 0$ and $x = l$. For a detailed explanation see the appendix at the end of the chapter.

Notes:

1. The convention for the sign of $u(x)$.
    If P moves to P′ in the direction of x increasing, i.e. B′P′ > BP (figure 3.2), then $u(x) > 0$. If P moves in the opposite direction, then $u(x) < 0$. This ensures that if, say, C is fixed and T_B is applied at B′, $u(0) < 0$, and then the term $T_B u(0)$ will give a negative contribution to the energy.

2. The functional (3.1) may be used even when the extensions $u(0)$ and $u(l)$ are known, as in chapter 2. In this case T_C and T_B are the reactions (unknown) which have produced the extensions (i.e. the forces which have been applied by the pins attaching the string).

The new functional (3.1) will be used to set up the finite element equations for the general case, which may be modified later to allow for particular boundary

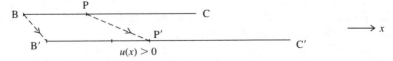

Figure 3.2   Convention for the sign of $u(x)$

conditions. If $u(0)$ and/or $u(l)$ are known, only a simple modification of the set of finite element equations will be required at the final stage to build in the given values.

3.3 Calculations element by element

It is this feature, that the calculations may be carried out 'element-by-element', which determines the strategy adopted by finite element computer programs. The theoretical basis for the process is the fundamental law 'that an integral may be evaluated by summing the values of the integral over subintervals which make up

Figure 3.3   Division of a domain into subdomains

the original interval' (figure 3.3). For example

$$\int_{x_1}^{x_N} f(x)\mathrm{d}x = \int_{x_1}^{x_2} f(x)\mathrm{d}x + \int_{x_2}^{x_3} f(x)\mathrm{d}x + \cdots + \int_{x_{N-1}}^{x_N} f(x)\mathrm{d}x$$

Suppose the interval $[0, l]$ of the model problem is divided into E elements of width $h = l/E$. (E is now used to denote the number of elements in the structure and hopefully will not be confused with the modulus of elasticity.)

Replacing the integrand by an $*$ as its precise form is not important at this stage, and dividing the interval into elements (figure 3.4), from (3.1)

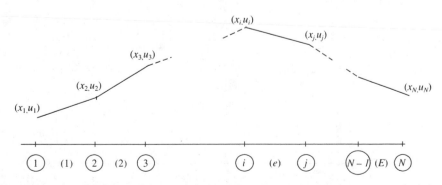

Figure 3.4   A discretisation and trial function

$$V(u(x)) = T_B u(0) + \int_0^l *\mathrm{d}x - T_C u(l)$$

$$= T_B u(0) + \underbrace{\int_{x_1=0}^{x_2} *\mathrm{d}x}_{V^1} + \underbrace{\int_{x_2}^{x_3} *\mathrm{d}x}_{V^2} + \cdots + \underbrace{\int_{x_{N-1}}^{x_N=l} *\mathrm{d}x - T_C u(l)}_{V^E}$$

$$= \sum_{e=1}^{E} V^e$$

where V^e is the energy of a typical element.

Note that in naming the integrals as energy, e.g. $V^2 = \int_{x_2}^{x_3} *\mathrm{d}x$, the work done by the tensions at the ends of the element has been left out. However, these tensions are internal forces and will cancel out with corresponding tensions from the neighbouring elements, so the combined result will not be in error. Thus we have the important result

$$V = \sum_{e=1}^{E} V^e \tag{3.2}$$

The total energy of the system is the sum of the energy of each element.

3.4 Calculations on a general element

The calculations can be worked out on a general element and the results used for each particular element of the system. A general element is shown in figure 3.5. (The node number j is, in fact, $i + 1$ in the context of a one-dimensional structure and we could use x_{i+1}, u_{i+1} instead of x_j, u_j, etc. The use of j gives a presentation which may be naturally extended to two or three dimensions.)

Allowing for the possibility of an external force at either end, the energy of the element is

$$V^e(u(x)) = T_i u(x_i) + \int_{x_i}^{x_j} \left[\left(\frac{\mathrm{d}u}{\mathrm{d}x} \right)^2 - (x + u) \right] \mathrm{d}x - T_j u(x_j) \tag{3.3}$$

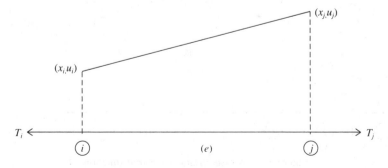

Figure 3.5   A general element

where the simplifying values $AE = 2$, $w = l = 1$ are used as in section 2.4. There are several new ideas involved in the calculations, so this section is subdivided.

3.4.1 Forming the approximation for u(x)

An assumption has to be made about how the extension $u(x)$ varies over each element. The simplest assumption is that it varies linearly. In this case combining the behaviour over each element gives a piecewise linear approximation over the whole string. The linear approximation was used in chapter 2 and will be mainly used in this book, although as previously mentioned it is possible and often desirable to make a higher order approximation. Most commonly, this is quadratic, which is likely to give more accurate results, but at the price of additional computation.

On the *e*th element, the approximation for $u(x)$ has the form

$$u^e(x) = \alpha + \beta x \tag{3.4}$$

where α and β are to be chosen so that

$$u^e(x_i) = u_i$$
$$\text{and} \quad u^e(x_j) = u_j,$$

and u_i and u_j are the unknown extensions at x_i and x_j, i.e.

$$u_i = \alpha + \beta x_i$$
$$u_j = \alpha + \beta x_j$$

Hence $\quad \alpha = \dfrac{u_i x_j - u_j x_i}{h}, \quad \beta = \dfrac{u_j - u_i}{h} \quad$ where $h = x_j - x_i$. $\tag{3.5}$

Thus (3.4) becomes $\quad u^e(x) = \dfrac{u_i x_j - u_j x_i}{h} + \dfrac{u_j - u_i}{h} x \tag{3.6}$

A more elegant and useful form of (3.6) may be found by introducing **shape functions**, $N_i(x)$ and $N_j(x)$ (figure 3.6). The graphs of the shape functions are straight lines and they have the fundamental properties

$$N_i(x) = \begin{cases} 1 & x = x_i \\ 0 & x = x_j \end{cases} \qquad N_j(x) = \begin{cases} 0 & x = x_i \\ 1 & x = x_j \end{cases} \tag{3.7}$$

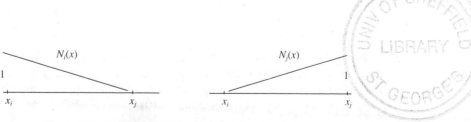

Figure 3.6   Linear, one-dimensional shape functions

It is straightforward to derive the expressions

$$N_i(x) = \frac{x_j - x}{h} \qquad N_j(x) = \frac{x - x_i}{h} \qquad (3.8)$$

Using these shape functions, an alternative form for the linear approximation for $u^e(x)$ becomes

$$u^e(x) = N_i(x)u_i + N_j(x)u_j \qquad (3.9)$$

It is left as an exercise for the reader to verify this by showing the equivalence of (3.6) and (3.9).

Exercise 3.1

(a) Show that N_i, N_j of (3.8) have the properties (3.7).

(b) Show that (3.6) and (3.9) are equivalent in two ways:

   (i) algebraically by manipulating (3.9) into the form (3.6),

   (ii) by observing that (3.9) is linear, and that it satisfies the properties $u^e(x_i) = u_i$ and $u^e(x_j) = u_j$ (this follows from the properties (3.7) and is the motivation for them).

Besides being elegant, the form (3.9) is capable of being extended to give higher order approximations, and may easily be generalised to two and three dimensions. Also, as the name implies, shape functions may be used in forming the geometry of elements. This will be discussed in chapter 8.

3.4.2 Energy of the element

Substituting (3.9) into (3.3) gives

$$
\begin{aligned}
V^e &= V^e(u^e(x)) \\
&= T_i u_i - T_j u_j \\
&\quad + \int_{x_i}^{x_j} \left[\left(N_i'(x)u_i + N_j'(x)u_j \right)^2 - x - N_i(x)u_i - N_j(x)u_j \right] dx \qquad (3.10)
\end{aligned}
$$

It is important to note that V^e is now a function of u_i and u_j, the unknown extensions at two nodes (the variable x will disappear after integrating between the limits x_i and x_j).

3.4.3 Differentiation

Both $\partial V^e / \partial u_i$ and $\partial V^e / \partial u_j$ will be needed for the next section. Theoretically, it does not matter whether the integration in (3.10) is carried out before the differentiation or after, but the manipulation is perhaps slightly easier if the differentiation comes first. Differentiating (3.10)

$$\frac{\partial V^e}{\partial u_i} = T_i + \int_{x_i}^{x_j} \left[2\Big(N_i'(x)u_i + N_j'(x)u_j\Big)N_i'(x) - N_i(x) \right] dx$$

using (3.8) and noting that $-N_i'(x) = N_j'(x) = \frac{1}{h}$

$$\frac{\partial V^e}{\partial u_i} = T_i + \int_{x_i}^{x_j} \left[\frac{2}{h^2}(u_i - u_j) - \frac{x_j - x}{h} \right] dx$$

$$= T_i + \frac{2}{h^2}(u_i - u_j)[x]_{x_i}^{x_j} + \left[\frac{(x_j - x)^2}{2h} \right]_{x_i}^{x_j}$$

giving

$$\frac{\partial V^e}{\partial u_i} = T_i + \frac{2}{h}(u_i - u_j) - \frac{h}{2} \tag{3.11}$$

Exercise 3.2
Show that

$$\frac{\partial V^e}{\partial u_j} = -T_j + \frac{2}{h}(-u_i + u_j) - \frac{h}{2} \tag{3.12}$$

The two equations, (3.11) and (3.12), provide the coefficients of the set of linear finite element equations which are discussed in the next section.

3.5 The finite element equations

These are the set of linear algebraic equations which arise from minimising the functional after the element assumption has been applied. Their solution is the finite element approximation to the extensions at the nodes. The derivation of the equations starts with the result (3.2), that the energy of the system is the sum of energy of each element

$$V = V^1 + V^2 + \cdots + V^E \tag{3.13}$$

When the linear approximation (3.9) is made for $u^e(x)$, the energy of each element is seen to be a function of u at its nodes – see (3.10). That is

$$V^1 = V^1(u_1, u_2), \quad V^2 = V^2(u_2, u_3), \cdots, V_E = V^E(u_{N-1}, u_N) \tag{3.14}$$

Thus the energy of the system has become a function of the nodal values of u

$$V = V(u_1, u_2, \cdots, u_N) \tag{3.15}$$

Originally, before any assumption concerning the form of the extension $u(x)$ was made, the search was for a function $u(x)$ to minimise V. Once the finite element trial function has been substituted for $u(x)$, V depends only on the nodal values of u, so the search for a curve becomes a search for values of $u_1, u_2, \cdots, u_N$ which minimise V. Thus $u_1, u_2, \cdots, u_N$ are chosen so that

$$\frac{\partial V}{\partial u_1} = \frac{\partial V}{\partial u_2} = \cdots = \frac{\partial V}{\partial u_N} = 0 \tag{3.16}$$

From (3.13)

$$\frac{\partial V}{\partial u_1} = 0 \Rightarrow \frac{\partial V^1}{\partial u_1} + \frac{\partial V^2}{\partial u_1} + \cdots + \frac{\partial V^E}{\partial u_1} = 0$$

$$\frac{\partial V}{\partial u_2} = 0 \Rightarrow \frac{\partial V^1}{\partial u_2} + \frac{\partial V^2}{\partial u_2} + \cdots + \frac{\partial V^E}{\partial u_2} = 0$$

$$\vdots \qquad\qquad \vdots \quad\ \vdots \quad \ddots \quad \vdots \quad\ \vdots$$

$$\frac{\partial V}{\partial u_N} = 0 \Rightarrow \frac{\partial V^1}{\partial u_N} + \frac{\partial V^2}{\partial u_N} + \cdots + \frac{\partial V^E}{\partial u_N} = 0 \tag{3.17}$$

These give rise to the finite element equations. When the equations (3.11) and (3.12) are used in (3.17), it is seen that they form a set of N linear equations in $u_1, u_2, \cdots, u_N$. This will become clearer when a specific example is considered.

3.6 Three-element solution

Suppose that the elements are chosen to be of equal size, $h = \frac{1}{3}$ (figure 3.7), then equations (3.11) and (3.12) for the general element (figure 3.8) become

$$\frac{\partial V^e}{\partial u_i} = T_i + 6(u_i - u_j) - \frac{1}{6} \tag{3.18}$$

$$\frac{\partial V^e}{\partial u_j} = 6(-u_i + u_j) - \frac{1}{6} - T_j \tag{3.19}$$

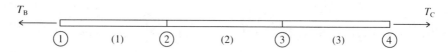

Figure 3.7   A three-element discretisation

Figure 3.8   The general element

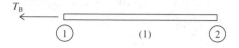

$$T_B \quad \leftarrow$$

①   (1)   ②

Figure 3.9   Element 1

The term $\partial V^1/\partial u_1$ can be obtained by relating element 1 (figure 3.9) to the general element and using equation (3.18). Thus element 1 $\leftrightarrow$ element e, and nodes 1 and 2 $\leftrightarrow$ nodes i and j, respectively. Also $T_B \leftrightarrow T_i$, thus (3.18) becomes

$$\frac{\partial V^1}{\partial u_1} = T_B + 6(u_1 - u_2) - \frac{1}{6}$$

It may be seen that V^2 and V^3 have no term in u_1, resulting in

$$\frac{\partial V^2}{\partial u_1} = \frac{\partial V^3}{\partial u_1} = 0$$

in other words, varying u_1 has no effect on the 'energy' of elements 2 and 3. Bringing the results together,

$$\frac{\partial V^1}{\partial u_1} + \frac{\partial V^2}{\partial u_1} + \frac{\partial V^3}{\partial u_1} = 0$$

becomes

$$\left(T_B + 6(u_1 - u_2) - \frac{1}{6} \right) + 0 + 0 = 0.$$

Now, when we consider varying u_2, the second equation of (3.17) is used, i.e.

$$\frac{\partial V}{\partial u_2} = 0 \Rightarrow \frac{\partial V^1}{\partial u_2} + \frac{\partial V^2}{\partial u_2} + \frac{\partial V^3}{\partial u_2} = 0$$

To find the expression for $\partial V^1/\partial u_2$ the element 1 must be related to the general element, as before, but it is the node corresponding to j which is now varying. So from (3.19),

$$\frac{\partial V^1}{\partial u_2} = 6(-u_1 + u_2) - \frac{1}{6}$$

To obtain $\partial V^2/\partial u_2$ the element 2 (figure 3.10) is related to the general element. Thus element 2 $\leftrightarrow$ element e, and nodes 2 and 3 $\leftrightarrow$ nodes i and j, respectively, and there are no external forces acting on the element. So from (3.18),

$$\frac{\partial V^2}{\partial u_2} = 6(u_2 - u_3) - \frac{1}{6}$$

②   (2)   ③

Figure 3.10   Element 2

Also, varying u_2 will not affect V^3, so $\partial V^3/\partial u_2 = 0$. Combining these results

$$\frac{\partial V^1}{\partial u_2} + \frac{\partial V^2}{\partial u_2} + \frac{\partial V^3}{\partial u_2} = 0$$

becomes

$$\left[6(-u_1 + u_2) - \frac{1}{6}\right] + \left[6(u_2 - u_3) - \frac{1}{6}\right] + 0 = 0$$

Exercise 3.3

Show that

$$\frac{\partial V^2}{\partial u_3} = 6(-u_2 + u_3) - \frac{1}{6},$$

$$\frac{\partial V^3}{\partial u_3} = 6(u_3 - u_4) - \frac{1}{6},$$

and that

$$\frac{\partial V^1}{\partial u_3} = 0$$

Hence show that

$$\frac{\partial V}{\partial u_3} = 0 \Rightarrow -6u_2 + 12u_3 - 6u_4 = \frac{1}{3}$$

Exercise 3.4

Show that

$$\frac{\partial V}{\partial u_4} = 0 \Rightarrow -6u_3 + 6u_4 = \frac{1}{6} + T_C$$

Thus, the finite element equations for u_1, u_2, u_3 and u_4 are

$$6u_1 - 6u_2 \qquad\qquad\qquad = \frac{1}{6} - T_B$$

$$-6u_1 + 12u_2 - 6u_3 \qquad = \frac{1}{3}$$

$$-6u_2 + 12u_3 - 6u_4 = \frac{1}{3}$$

$$-6u_3 + 6u_4 = \frac{1}{6} + T_C \qquad\qquad (3.20)$$

or, briefly

$$\mathbf{Ku = f} \qquad\qquad (3.21)$$

where

$$\mathbf{K} = \begin{bmatrix} 6 & -6 & 0 & 0 \\ -6 & 12 & -6 & 0 \\ 0 & -6 & 12 & -6 \\ 0 & 0 & -6 & 6 \end{bmatrix} \qquad\qquad (3.22)$$

is called the **global stiffness matrix**, $\mathbf{u} = [u_1, u_2, u_3, u_4]^{\mathrm{T}}$ is the column vector of the unknown extensions, and

$$\mathbf{f} = \left[\frac{1}{6} - T_B, \frac{1}{3}, \frac{1}{3}, \frac{1}{6} + T_C\right]^{\mathrm{T}} \tag{3.23}$$

is called the **global force vector**.

It can be seen that the matrix $\mathbf{K}$ is singular as the columns all sum to zero, indicating that $\det \mathbf{K} = 0$. So the system of equations will have no solution unless the right-hand sides also sum to zero, in which case there will be infinitely many solutions. If the elements of $\mathbf{f}$ sum to zero, then $T_B = T_C + 1$, which is just the requirement that the string should be in equilibrium under the external forces and its own weight $(wl = 1)$. In this case one displacement may be chosen arbitrarily and the others solved relative to it – say choose $u_1 = 0$ and solve for u_2, u_3 and u_4, which will then be the displacements at nodes 2, 3 and 4 relative to node 1.

Exercise 3.5
If $T_B = 2$ and $T_C = 1$ solve the equations (3.20).

The set of equations (3.20) may be modified to incorporate the other types of boundary conditions listed in section 3.1. Consider for example B′ fixed and C extended by a known force $T_C = 1$. Since B′ is fixed $u(0) = u_1 = 0$. Now the first equation of the set (3.20) comes from $\partial V/\partial u_1 = 0$, i.e. from varying u_1, but as u_1 is fixed the equation is not relevant and may be left out at this stage. The remaining equations, together with $u_1 = 0$ and $T_C = 1$, become

$$12u_2 - 6u_3 \qquad\quad = \frac{1}{3}$$

$$-6u_2 + 12u_3 - 6u_4 = \frac{1}{3}$$

$$-6u_3 + 6u_4 = \frac{7}{6} + 1$$

which give

$$u_2 = \frac{11}{36}, \qquad u_3 = \frac{20}{36}, \qquad u_4 = \frac{27}{36}$$

Note that the first equation of (3.20),

$$6u_1 - 6u_2 = \frac{1}{6} - T_B$$

while not having been used, is still true. If the values for u_1 and u_2 are substituted,

$$0 - 6 \times \frac{11}{36} = \frac{1}{6} - T_B$$

i.e. $\tag{3.24}$

$$T_B = 2$$

which is the reaction at the point where the string is attached.

Exercise 3.6
Solve (3.20) where $u(0) = 0$ and $u(1) = 1$ and compare with the analytical solution $u = \frac{x}{4}(5 - x)$

Exercise 3.7
Suppose the string is attached at B′ and hangs freely under its own weight.

(a) What is the functional that has to be minimised?

(b) Modify and solve the equations (3.20).

(c) Show the finite element solution on a graph, and also the tension $T = 2\mathrm{d}u/\mathrm{d}x$ (see section 2.5).

Examine the value of T at $x = 1$. [The analytical solution is $u = \frac{x}{4}(2 - x)$ and $T = (1 - x)$.]

Appendix

Derivation of the functional for elasticity
The finite element method for problems in elasticity is based on the *Theorem of Potential Energy* which states: 'Of all displacements satisfying the given geometric boundary conditions, those that satisfy the equations of static equilibrium are distinguished by giving a stationary value of the potential energy. If the stationary condition is a minimum, the equilibrium is stable.'

The total potential energy V consists of two parts

$$V = V_{\mathrm{s}} + V_{\mathrm{a}} \tag{A.1}$$

where V_{s} is the strain energy from the effect of internal forces, together with the gravitational potential energy which we derived in chapter 2, and V_{a} is the potential energy associated with the applied external forces. For a conservative system the loss of potential energy during the loading process owing to the action of the external forces must be equal to the work done, W_{a}, on the system by the external forces, i.e. $-V_{\mathrm{a}} = W_{\mathrm{a}}$. In calculating the work done by a force, force × displacement is used.

Note that the force is always assumed to act at full intensity and its potential energy arises from its magnitude and its capacity to cause displacements. In our example, where the forces are applied at both ends of the string,

$$W_{\mathrm{a}} = T_C u(l) - T_B u(0)$$

Also note that if $u(0) < 0$, as will often be the case, then the work done by T_B will be positive. Putting $V_{\mathrm{a}} = -W_{\mathrm{a}}$ in (A.1) gives the required equation (3.1).

General exercises for chapter 3

1. Obtain the four-element solution to the model problem with the boundary conditions $u(0) = 0$ and $T_C = 1$.

2. Consider the variational problem: find $u(x)$ on the interval $[0, 1]$ such that $u(0) = 0$ and u minimises

$$\int_0^1 \left(u'(x)^2 - u(x) \right) dx - u(1).$$

   Obtain a three-element solution.

3. Solve the model problem with $u(0) = 0$ and $u(1) = 1$ as in exercise 3.6. In this exercise subdivide the interval $[0, 1]$ into 3 elements of width 1/4, 1/4, 1/2 and again compare with the analytical solution $u = \frac{x}{4}(5 - x)$. (Usually, choosing smaller elements in a region leads to a more accurate solution there, but in this case the finite element values for u are exact anyway.)

   In exercises 4, 6 and 7, first find the element equations corresponding to (3.11) and (3.12) and then use them to assemble the global equations in the form of (3.21).

4. Use four equal elements to minimise

$$V(u(x)) = \int_0^1 \left[\left(\frac{du}{dx} \right)^2 + xu \right] dx$$

   where $u(0) = 0$, $u(1) = 0$.

5. As a preliminary to exercise 6, show that

$$\int_{x_i}^{x_j} N_i^2(x)dx = \int_{x_i}^{x_j} N_j^2(x)dx = \frac{h}{3}, \quad \int_{x_i}^{x_j} N_i(x)N_j(x)dx = \frac{h}{6}$$

6. Use four equal elements to minimise

$$V(u(0)) = \int_0^1 \left[\left(\frac{du}{dx} \right)^2 + u^2 \right] dx$$

   where $u(0) = 1$.

7. Use three linear elements to approximate to u satisfying $u(0) = 0$, and minimising

$$V(u) = \int_0^1 \frac{1}{2} \left[\left(\frac{du}{dx} \right)^2 + u^2 - 2u \right] dx + u(1).$$

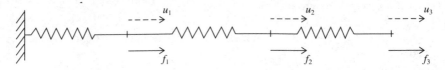

Figure 3.11   A three-spring system

8. A three-spring system is attached at A and lies horizontally (figure 3.11). The springs, of stiffness k_1, k_2, k_3, are stretched as shown by the three forces f_1, f_2 and f_3 causing the displacements u_1, u_2 and u_3. When $u_1 = u_2 = u_3 = 0$ the springs are unstretched.

Write down the potential energy of the system in terms of the spring stiffness and the fs and us and show that the finite element equations give the matrix equation:

$$\begin{bmatrix} k_1 + k_2 & -k_2 & 0 \\ -k_2 & k_2 + k_3 & -k_3 \\ 0 & -k_3 & k_3 \end{bmatrix} \begin{bmatrix} u_1 \\ u_2 \\ u_3 \end{bmatrix} = \begin{bmatrix} f_1 \\ f_2 \\ f_3 \end{bmatrix}$$

Solve the equation when $k_1 = k_2 = k_3 = 1$ and $f_1 = f_2 = f_3 = 4$.

[The potential energy gained by a spring when being extended a distance u is $\frac{1}{2} k u^2$.]

4 The matrix approach

4.1 Introduction

It is not a coincidence that the finite element method blossomed at the time when computers were being developed, because it calls for a vast amount of computation for a problem of even modest size. It requires the assembly and solution of a large system of linear equations and, in addition, benefits from the computer's ability to present data graphically both to verify the problem structure and to display the solution. This chapter is about the way the computer organises the method and needs the language of matrices.

The statement that 'the global stiffness matrix is the sum of the element stiffness matrices' describes the process at the heart of the computer generation of the finite element equations. However, besides justifying the basic process, some detailed explanation is necessary because, for one thing, the global and element matrices would seem to have different dimensions. The chapter contains:

- The idea of a condensed and full-sized matrix and how it is used in obtaining the above statement. The matrix description of the method is established.

- An explanation of the computer (and by hand) method for building up the global matrices from the element matrices in condensed form, together with a technique for imposing boundary conditions.

4.2 Casting the element equations in matrix form

Consider again the general element of section 3.4 (figure 4.1).

The following results were obtained in section 3.4.3 and stated as equations (3.11) and (3.12)

$$\frac{\partial V^e}{\partial u_i} = T_i + \frac{2}{h}(u_i - u_j) - \frac{h}{2} \tag{4.1}$$

$$\frac{\partial V^e}{\partial u_j} = -T_j + \frac{2}{h}(-u_i + u_j) - \frac{h}{2} \tag{4.2}$$

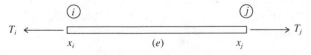

Figure 4.1 A general element

These equations contain the information which is contributed by the element to the global finite element equations (3.21). To enter into the theme of the chapter the equations are written in matrix form:

$$
\begin{bmatrix} \dfrac{\partial V^e}{\partial u_i} \\ \dfrac{\partial V^e}{\partial u_j} \end{bmatrix} = \begin{bmatrix} \dfrac{2}{h} & -\dfrac{2}{h} \\ -\dfrac{2}{h} & \dfrac{2}{h} \end{bmatrix} \begin{bmatrix} u_i \\ u_j \end{bmatrix} - \begin{bmatrix} \dfrac{h}{2} - T_i \\ \dfrac{h}{2} + T_j \end{bmatrix}
\tag{4.3}
$$

This may be written briefly, by introducing symbols for each of the matrices, as

$$
\frac{\partial V^e}{\partial \mathbf{u}^e} = \mathbf{K}_c^e \mathbf{u}^e - \mathbf{f}_c^e
\tag{4.4}
$$

where the vector of the derivatives of V with respect to the element unknowns

$$
\begin{bmatrix} \dfrac{\partial V^e}{\partial u_i} \\ \dfrac{\partial V^e}{\partial u_j} \end{bmatrix} \quad \text{is written as} \quad \frac{\partial V^e}{\partial \mathbf{u}^e}
$$

The coefficient matrix of the element unknowns, the element stiffness matrix,

$$
\begin{bmatrix} \dfrac{2}{h} & -\dfrac{2}{h} \\ -\dfrac{2}{h} & \dfrac{2}{h} \end{bmatrix} \quad \text{is written as} \quad \mathbf{K}_c^e.
$$

The subscript 'c' refers to the matrix being in a condensed form rather than in the expanded form which will be introduced later in the chapter. The vector of the unknowns present in element e

$$
\begin{bmatrix} u_i \\ u_j \end{bmatrix} \quad \text{is written as} \quad \mathbf{u}^e, \text{ and } \begin{bmatrix} \frac{2}{h} - T_i \\ \frac{2}{h} + T_j \end{bmatrix} \quad \text{as} \quad \mathbf{f}_c^e.
$$

To illustrate with the three-element division of the problem domain considered in section 3.6, the general form (4.3) becomes for element 1,

$$
\begin{bmatrix} \dfrac{\partial V^1}{\partial u_1} \\ \dfrac{\partial V^1}{\partial u_2} \end{bmatrix} = \begin{bmatrix} \dfrac{2}{h} & -\dfrac{h}{2} \\ -\dfrac{2}{h} & \dfrac{h}{2} \end{bmatrix} \begin{bmatrix} u_1 \\ u_2 \end{bmatrix} - \begin{bmatrix} \dfrac{h}{2} - T_B \\ \dfrac{h}{2} \end{bmatrix}
\tag{4.5}
$$

In order to develop a notation that can express in fairly simple language the formation of the finite element equations $\mathbf{Ku} = \mathbf{f}$, it is necessary to expand the two equations of (4.5) to include contributions from differentiating V^1 with respect to *all* the unknowns. It was shown previously in section 3.5 that V^1 depends only on u_1 and u_2, so that

$$
\frac{\partial V^1}{\partial u_3} = \frac{\partial V^1}{\partial u_4} = 0
\tag{4.6}
$$

The two equations contained in (4.6) are now combined with those of (4.5). In doing so the vector of element unknowns $\mathbf{u}^1 = [u_1 \ u_2]^T$ is expanded to include all the unknowns of the problem, $\mathbf{u} = [u_1 \ u_2 \ u_3 \ u_4]^T$

$$
\begin{bmatrix} \frac{\partial V^1}{\partial u_1} \\ \frac{\partial V^1}{\partial u_2} \\ \frac{\partial V^1}{\partial u_3} \\ \frac{\partial V^1}{\partial u_4} \end{bmatrix} = \begin{bmatrix} \frac{2}{h} & -\frac{2}{h} & 0 & 0 \\ -\frac{2}{h} & \frac{2}{h} & 0 & 0 \\ 0 & 0 & 0 & 0 \\ 0 & 0 & 0 & 0 \end{bmatrix} \begin{bmatrix} u_1 \\ u_2 \\ u_3 \\ u_4 \end{bmatrix} - \begin{bmatrix} \frac{h}{2} - T_B \\ \frac{h}{2} \\ 0 \\ 0 \end{bmatrix}
\tag{4.7}
$$

Exercise 4.1
Satisfy yourself that the matrix equation (4.7) combines (4.5) and (4.6).

Following the notation of equation (4.4), (4.7) is written as

$$
\frac{\partial V^1}{\partial \mathbf{u}} = \mathbf{K}^1 \mathbf{u} - \mathbf{f}^1
\tag{4.8}
$$

Note that in (4.7), the previous matrices $\mathbf{K}_c^1$ and $\mathbf{f}_c^1$ have been expanded by including zeros in appropriate places to become $\mathbf{K}^1$ and $\mathbf{f}^1$. The matrices now relate to all the unknowns and are no longer referred to as 'condensed'. $\mathbf{K}^1$ and $\mathbf{f}^1$ are the *stiffness matrix* and *force vector* of element 1.

In a similar way, the equation (4.3) when applied to element 2 is

$$
\begin{bmatrix} \frac{\partial V^2}{\partial u_2} \\ \frac{\partial V^2}{\partial u_3} \end{bmatrix} = \begin{bmatrix} \frac{2}{h} & -\frac{2}{h} \\ -\frac{2}{h} & \frac{2}{h} \end{bmatrix} \begin{bmatrix} u_2 \\ u_3 \end{bmatrix} - \begin{bmatrix} \frac{h}{2} \\ \frac{h}{2} \end{bmatrix}
\tag{4.9}
$$

and when expanded becomes

$$
\begin{bmatrix} \frac{\partial V^2}{\partial u_1} \\ \frac{\partial V^2}{\partial u_2} \\ \frac{\partial V^2}{\partial u_3} \\ \frac{\partial V^2}{\partial u_4} \end{bmatrix} = \begin{bmatrix} 0 & 0 & 0 & 0 \\ 0 & \frac{2}{h} & -\frac{2}{h} & 0 \\ 0 & -\frac{2}{h} & \frac{2}{h} & 0 \\ 0 & 0 & 0 & 0 \end{bmatrix} \begin{bmatrix} u_1 \\ u_2 \\ u_3 \\ u_4 \end{bmatrix} - \begin{bmatrix} 0 \\ \frac{h}{2} \\ \frac{h}{2} \\ 0 \end{bmatrix}
\tag{4.10}
$$

or, briefly

$$
\frac{\partial V^2}{\partial \mathbf{u}} = \mathbf{K}^2 \mathbf{u} - \mathbf{f}^2
\tag{4.11}
$$

Exercise 4.2
Write down the matrices $\mathbf{K}^3$ and $\mathbf{f}^3$

4.3 The finite element equations

The global set of finite element equations for the discretisation into three elements are, as explained in section 3.5,

$$\frac{\partial V}{\partial u_1} = 0 \quad \Rightarrow \quad \frac{\partial V^1}{\partial u_1} + \frac{\partial V^2}{\partial u_1} + \frac{\partial V^3}{\partial u_1} = 0$$

$$\frac{\partial V}{\partial u_2} = 0 \quad \Rightarrow \quad \frac{\partial V^1}{\partial u_2} + \frac{\partial V^2}{\partial u_2} + \frac{\partial V^3}{\partial u_2} = 0$$

$$\frac{\partial V}{\partial u_3} = 0 \quad \Rightarrow \quad \frac{\partial V^1}{\partial u_3} + \frac{\partial V^2}{\partial u_3} + \frac{\partial V^3}{\partial u_3} = 0$$

$$\frac{\partial V}{\partial u_4} = 0 \quad \Rightarrow \quad \frac{\partial V^1}{\partial u_4} + \frac{\partial V^2}{\partial u_4} + \frac{\partial V^3}{\partial u_4} = 0 \qquad (4.12)$$

Writing

$$\left[\frac{\partial V^1}{\partial u_1} \ \frac{\partial V^1}{\partial u_2} \ \frac{\partial V^1}{\partial u_3} \ \frac{\partial V^1}{\partial u_4} \right]^{\mathrm{T}} \quad \text{as} \quad \frac{\partial V}{\partial \mathbf{u}}, \text{etc.}$$

the four equations of (4.12) can be compressed into one matrix equation

$$\frac{\partial V^1}{\partial \mathbf{u}} + \frac{\partial V^2}{\partial \mathbf{u}} + \frac{\partial V^3}{\partial \mathbf{u}} = \mathbf{0} \qquad (4.13)$$

Using the results of section 4.1, this becomes

$$\mathbf{K}^1 \mathbf{u} - \mathbf{f}^1 + \mathbf{K}^2 \mathbf{u} - \mathbf{f}^2 + \mathbf{K}^3 \mathbf{u} - \mathbf{f}^3 = \mathbf{0}$$

i.e.

$$\left[\mathbf{K}^1 + \mathbf{K}^2 + \mathbf{K}^3 \right] \mathbf{u} = \mathbf{f}^1 + \mathbf{f}^2 + \mathbf{f}^3$$

or

$$\mathbf{K} \mathbf{u} = \mathbf{f} \qquad (4.14)$$

where

$$\mathbf{K} = \sum_{e=1}^{3} \mathbf{K}^e \text{ is the global stiffness matrix} \qquad (4.15)$$

and

$$\mathbf{f} = \sum_{e=1}^{3} \mathbf{f}^e \text{ is the global force vector} \qquad (4.16)$$

Exercise 4.3

Verify by matrix addition that (4.15) and (4.16) give the same result as that shown in (3.22) and (3.23) (where $h = \frac{1}{3}$).

4.4 Forming the global matrix

In forming the global stiffness matrix as the sum of the element stiffness matrices it is clearly not necessary to carry out the expansion of the condensed matrices before they are added. All that is needed is to work out the appropriate positions in the global matrix which correspond to the particular condensed element matrix and then just add in the numbers. The process is part of the general finite element method and is known as **assembling** the element matrices; it is simple for the one-dimensional model problem that we are considering.

Consider a general subdivision into E elements and $N = E + 1$ nodes, so that $h = 1/E$. The condensed matrix for element 1 is

$$
\begin{array}{c}
 \\
\frac{\partial V^1}{\partial u_1} \\
\frac{\partial V^1}{\partial u_2}
\end{array}
\begin{array}{cc}
u_1 & u_2
\end{array}
\left[
\begin{array}{cc}
\frac{2}{h} & -\frac{2}{h} \\
-\frac{2}{h} & \frac{2}{h}
\end{array}
\right]
$$

where on the border of the matrix the appropriate rows and columns have been indicated. Suppose we start with an empty shell for the global matrix and add in the contribution from the first element. This gives

$$
\begin{array}{c}
\frac{\partial V}{\partial u_1} \\
\frac{\partial V}{\partial u_2} \\
\frac{\partial V}{\partial u_3} \\
\frac{\partial V}{\partial u_4} \\
\vdots \\
\frac{\partial V}{\partial u_N}
\end{array}
\begin{array}{cccccc}
u_1 & u_2 & u_3 & u_4 & \cdots & u_N
\end{array}
\left[
\begin{array}{cccccc}
\frac{2}{h} & -\frac{2}{h} & \cdot & \cdot & \cdots & \cdot \\
-\frac{2}{h} & \frac{2}{h} & \cdot & \cdot & \cdots & \cdot \\
\cdot & \cdot & \cdot & \cdot & \cdots & \cdot \\
\cdot & \cdot & \cdot & \cdot & \cdots & \cdot \\
\vdots & \vdots & \vdots & \vdots & \ddots & \vdots \\
\cdot & \cdot & \cdot & \cdot & \cdots & \cdot
\end{array}
\right]
$$

For element 2 the condensed matrix is

$$
\begin{array}{c}
 \\
\frac{\partial V^2}{\partial u_2} \\
\frac{\partial V^2}{\partial u_3}
\end{array}
\begin{array}{cc}
u_2 & u_3
\end{array}
\left[
\begin{array}{cc}
\frac{2}{h} & -\frac{2}{h} \\
-\frac{2}{h} & \frac{2}{h}
\end{array}
\right]
$$

adding this into the current form of the global matrix gives

$$
\begin{array}{c}
\frac{\partial V}{\partial u_1} \\
\frac{\partial V}{\partial u_2} \\
\frac{\partial V}{\partial u_3} \\
\frac{\partial V}{\partial u_4} \\
\vdots \\
\frac{\partial V}{\partial u_N}
\end{array}
\begin{array}{cccccc}
u_1 & u_2 & u_3 & u_4 & \cdots & u_N
\end{array}
\left[
\begin{array}{cccccc}
\frac{2}{h} & -\frac{2}{h} & \cdot & \cdot & \cdots & \cdot \\
-\frac{2}{h} & \frac{4}{h} & -\frac{2}{h} & \cdot & \cdots & \cdot \\
\cdot & -\frac{2}{h} & \frac{2}{h} & \cdot & \cdots & \cdot \\
\cdot & \cdot & \cdot & \cdot & \cdots & \cdot \\
\vdots & \vdots & \vdots & \vdots & \ddots & \vdots \\
\cdot & \cdot & \cdot & \cdot & \cdots & \cdot
\end{array}
\right]
$$

Note that the term in position (2,2) contains the sum of the contributions from both element 1 and 2. It is left as an exercise to complete the pattern.

Exercise 4.4
Write down the general stiffness matrix started in section 4.4.

Exercise 4.5
Form the general force vector **f**.

Exercise 4.6
Obtain **K** and **f** for the five-element subdivision (of unequal size) shown in figure 4.2.

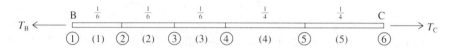

Figure 4.2   A non-uniform discretisation

4.5 Applying the boundary conditions

The global set of finite element equations for the division into three equal elements have been derived in chapter 3, equation (3.20). They are, before the boundary conditions have been applied,

$$
\begin{bmatrix}
6 & -6 & 0 & 0 \\
-6 & 12 & -6 & 0 \\
0 & -6 & 12 & -6 \\
0 & 0 & -6 & 6
\end{bmatrix}
\begin{bmatrix}
u_1 \\
u_2 \\
u_3 \\
u_4
\end{bmatrix}
=
\begin{bmatrix}
\frac{1}{6} - T_B \\
\frac{1}{3} \\
\frac{1}{3} \\
\frac{1}{6} + T_C
\end{bmatrix}
$$

Consider the stiffness matrix **K**. Besides being singular as discussed in section 3.6, it is also seen to be symmetrical (about the leading diagonal). This is usually so in a finite element analysis, and for problems in linear elasticity and in heat flow it is always the case. Also, **K** has a banded structure, in that the non-zero terms lie in a band symmetrically placed about the leading diagonal.

Computer programs implementing the finite element method take advantage of this banded structure and symmetry of **K** to store only half the band. The dimension of **K** can be large, even for simple problems, so that it is useful to be able to reduce the storage requirements.

Exercise 4.7

   (a) What is the bandwidth of **K** when (i) 4 elements, (ii) 5 elements and (iii) E elements are used?

   (b) For the general case of E elements (i) how many numbers are in **K**, (ii) how many of these are non-zero and (iii) how many need to be stored in the computer?

Consider how boundary conditions may be imposed through a computer program. Boundary conditions are classified into two types:

1. **Essential** boundary conditions, where the value of the problem variable itself is known. These are 'built in' to the solution and consequently are satisfied exactly.

2. **Natural** boundary conditions, which involve the derivative of the unknown, and are introduced through the functional. In the model problem the natural boundary condition is to prescribe the tension $T = AEdu/dx$. In general, the finite element solution only approximates to this type of boundary condition.

How these two types of condition may be imposed 'by hand' has been illustrated in section 3.6. Now, a similar example is considered to show how they may be incorporated by a computer program. There is not a unique way of doing this, but the method shown is in common use.

Suppose that the (somewhat unlikely) boundary conditions are $u(0) = -1$, i.e. $u_1 = -1$, and $T_C = 2$. Before these are applied, the equations are

$$6u_1 - 6u_2 \qquad\qquad = \frac{1}{6} - T_B$$

$$-6u_1 + 12u_2 - 6u_3 \qquad = \frac{1}{3}$$

$$-6u_2 + 12u_3 - 6u_4 = \frac{1}{3}$$

$$-6u_3 + 6u_4 = \frac{1}{6} + T_C \qquad\qquad (4.17)$$

The first equation is not needed initially, as T_B is unknown. It is replaced by another equation whose solution gives the prescribed value $u_1 = -1$. For convenience this is $6u_1 = -6$, which can be obtained by simply zeroing the off-diagonal terms of the first row of **K** and replacing the first term of **f** by -6. At this stage the equations are, on including the given value for T_C,

$$6u_1 \qquad\qquad\qquad = -6$$

$$-6u_1 + 12u_2 - 6u_3 \qquad = \frac{1}{3}$$

$$-6u_2 + 12u_3 - 6u_4 = \frac{1}{3}$$

$$-6u_3 + 6u_4 = \frac{1}{6} + 2$$

The coefficient matrix **K** has kept the same dimensions but is no longer symmetrical, so an advantage has been lost. However it can be regained if the '$-6u_1$' of row 2 is replaced by zero. This can be done by using a Gaussian elimination operation, i.e. by adding row 1 to row 2, giving

$$6u_1 \qquad\qquad\qquad\qquad = -6$$

$$12u_2 - 6u_3 \qquad\qquad = \frac{1}{3} - 6$$

$$-6u_2 + 12u_3 - 6u_4 = \frac{1}{3}$$

$$-6u_3 + 6u_4 = \frac{1}{6} + 2$$

The coefficient matrix

$$\begin{bmatrix} 6 & 0 & 0 & 0 \\ 0 & 12 & -6 & 0 \\ 0 & -6 & 12 & -6 \\ 0 & 0 & -6 & 6 \end{bmatrix}$$

is again symmetrical. Solving the equations gives

$$u_1 = -1, \qquad u_2 = -\frac{19}{36}, \qquad u_3 = -\frac{4}{36}, \qquad u_4 = \frac{9}{36}$$

4.6 Displaying the solution

The finite element approximation to the displacement is displayed, together with the corresponding tension, in figure 4.3.

Observe from the graph that the essential boundary condition $u(0) = -1$ is reproduced exactly by the finite element solution (as it must be since it was incorporated into the solution), but the tension at $x = 1$, $T = 2$, is only approximately satisfied: the value given is

$$2\frac{du}{dx} = \frac{13}{6} = 2.17\ldots$$

and not the prescribed value.

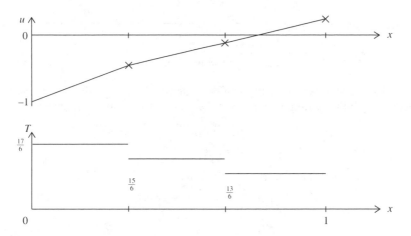

Figure 4.3   The finite element solution for displacement and tension

---

Exercise 4.8
Solve (4.17) with the boundary conditions
$$T_B = 2, \qquad u(1) = 0$$
incorporating them in the manner described for the computer method.

Draw a graph of the finite element solution for u and T, and consider how well the boundary conditions are satisfied.

4.7 An axisymmetric problem in heat flow

Consider the heat flow in a long thick hollow cylinder of inner and outer radii a and b, with no internal heat source (figure 4.4). The partial differential equation satisfied by temperature, u, is Laplace's equation which, when expressed in cylindrical coordinates (r, θ, z), is

$$k\left[\frac{\partial^2 u}{\partial r^2} + \frac{1}{r}\frac{\partial u}{\partial r} + \frac{1}{r^2}\frac{\partial^2 u}{\partial \theta^2} + \frac{\partial^2 u}{\partial z^2}\right] = 0$$

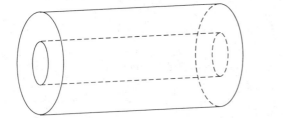
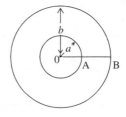

Figure 4.4   A hollow cylinder with thick walls

Finite Elements

Figure 4.5   The problem domain

where k is the thermal conductivity of the material. If the cylinder is long, and the boundary conditions do not vary along the length, the temperature is independent of z. So

$$\frac{\partial u}{\partial z} = \frac{\partial^2 u}{\partial z^2} = 0$$

then the problem reduces to two dimensions

$$k\left[\frac{\partial^2 u}{\partial r^2} + \frac{1}{r}\frac{\partial u}{\partial r} + \frac{1}{r^2}\frac{\partial^2 u}{\partial \theta^2}\right] = 0$$

Further, if the boundary conditions are axisymmetric, then u will be independent of θ and may be found by solving

$$k\left[\frac{d^2 u}{dr^2} + \frac{1}{r}\frac{du}{dr}\right] = 0 \tag{4.18}$$

in the one-dimensional region $a < r < b$ shown in figure 4.5, where a and b are the internal and external radii of the cylinder. The possible types of boundary conditions are that:

1. u is prescribed,
2. the heat flux, $q = -k\dfrac{du}{dr}$, is prescribed, or
3. the heat flux on the boundary is transferred by convection so that $q = h(u - u_\infty)$ where h is the convection coefficient and u_∞ is the temperature of the fluid outside (or inside) the cylinder. This condition will be considered later in chapter 5 for two-dimensional heat flow, but not now.

As in the problem of the elastic string, the problem variable, u either satisfies a differential equation or minimises a functional. The derivation of the variation form from the differential equation is the subject of chapters 7 and 8: here we will quote the functional and leave any curiosity to be satisfied later. For the general case where the heat fluxes at the boundaries are q_A and q_B, and assuming that there is no internal heat source, the temperature minimises the functional

$$V(u(r)) = \int_a^b \frac{1}{2} kr\left(\frac{du}{dr}\right)^2 dr + au_A q_A + bu_B q_B \tag{4.19}$$

where the direction of the heat fluxes are shown in figure 4.6.

Figure 4.6   Boundary heat fluxes

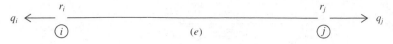

Figure 4.7 The general element

4.7.1 Calculations for a general element

The functional (4.19) becomes for the element e (shown in figure 4.7)

$$V^e(u) = \int_{r_i}^{r_j} \frac{1}{2} kr \left(\frac{du}{dr}\right)^2 dr + r_i u_i q_i + r_j u_j q_j \tag{4.20}$$

The next stage is to make the linear approximation for u on the element

$$u^e = N_i(r)u_i + N_j(r)u_j$$

and substitute this into the functional (4.20). The derivation of the element equations and further calculations are left as exercises.

Exercise 4.9
Show that

$$\frac{\partial V^e}{\partial u_i} = \frac{k}{2h}(r_j + r_i)(u_i - u_j) + r_i q_i$$

$$\frac{\partial V^e}{\partial u_j} = \frac{k}{2h}(r_j + r_i)(-u_i + u_j) + r_j q_j \tag{4.21}$$

where $h = r_j - r_i$, and hence that

$$\mathbf{K}_c^e = \frac{k}{2h}(r_j + r_i) \begin{bmatrix} 1 & -1 \\ -1 & 1 \end{bmatrix}$$

and

$$\mathbf{f}_c^e = -\begin{bmatrix} r_i q_i \\ r_j q_j \end{bmatrix}.$$

4.7.2 Forming the finite element equations

Exercise 4.10
For the particular values, $a = 1$, $b = 2$ and with a three equal element subdivision (figure 4.8), obtain the finite element equations

$$\mathbf{Ku} = \mathbf{f}$$

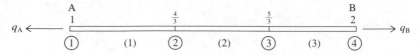

Figure 4.8   A three-element discretisation

where
$$\mathbf{K} = \frac{k}{2}\begin{bmatrix} 7 & -7 & 0 & 0 \\ -7 & 16 & -9 & 0 \\ 0 & -9 & 20 & -11 \\ 0 & 0 & -11 & 11 \end{bmatrix}$$

and
$$\mathbf{u} = [u_1\, u_2\, u_3\, u_4]^{\mathrm{T}}, \qquad \mathbf{f} = [-q_{\mathrm{A}}\, 0\, 0\, -2q_{\mathrm{B}}]^{\mathrm{T}}$$

4.7.3 Various boundary conditions

Exercise 4.11
For exercise 4.10, and putting $k = 1$ for simplicity, solve with the following boundary conditions, comparing graphically the finite element solution with the analytical solution for both the temperature u and heat flux $q(r) = -k(\mathrm{d}u/\mathrm{d}r)$

(a) $q_{\mathrm{A}} = -10$ an input of heat, $u(2) = 0$

(b) $u(1) = 0$, $u(2) = 100$

[The exact solutions are (a) $u(r) = -10\ln(\frac{r}{2})$ (2) $u(r) = 100(\ln r/\ln 2)$.]

Exercise 4.12
Deduce from the finite element equations that if both q_{A} and q_{B} are prescribed, a solution exists only if

$$r_{\mathrm{A}}q_{\mathrm{A}} + r_{\mathrm{B}}q_{\mathrm{B}} = 0$$

This is the conservation law for flux. The solution in this case is unique to within a constant.

General exercises for chapter 4

1.   Consider the variational problem of chapter 3, general exercise 4.

$$V(u(x)) = \int_0^1 \left[\left(\frac{\mathrm{d}u}{\mathrm{d}x}\right)^2 + xu \right] \mathrm{d}x$$

where $u(0) = 0$, $u(1) = 0$. Obtain the general element's condensed stiffness matrix and force vector (again reference may be made to chapter 3). Form the global set of finite element equations for E equal elements and N nodes using the method of section 4.4.

2. Repeat the work of exercise 1, using (the functional of chapter 3, general exercise 4)

$$V(u(x)) = \int_0^1 \left[\left(\frac{du}{dx} \right)^2 + u^2 \right] dx$$

where $u(0) = 1$.

3. A conducting rod AB, of unit length, has a prescribed temperature at the end A and loses heat through convection at B (figure 4.9). For convenience both k and h are chosen to be 1. Expressed in variational form the temperature u satisfies $u = 100$ at A and minimises

$$V(u(x)) = \int_0^1 \frac{1}{2} \left(\frac{du}{dx} \right)^2 dx + (u_B - 40)^2$$

where u_B is the unknown temperature at B.

A   B

$u = 100$ u_B

Figure 4.9   A conducting rod

Derive the stiffness matrix and force vector for a two-noded linear element. Hence form the finite element equations for four equally sized elements and obtain their solution. Show, using graphs, the temperature and heat flow $q = -(du/dx)$. How much heat must be input at A in order to maintain the temperature at 100 °C?

4. Use linear elements to approximate to u satisfying $u(0) = 0$, and minimising

$$V(u) = \int_0^1 \frac{1}{2} \left[\left(\frac{du}{dx} \right)^2 + u^2 - 2u \right] dx + u(1)$$

5. An insulating wall is constructed of three homogeneous layers of differing conductivity, $k_1 = 5, k_2 = 3, k_3 = 7$, and widths $w_1 = 0.05$, $w_2 = 0.06$, $w_3 = 0.02$ (figure 4.10). At the inner side A the temperature is given as

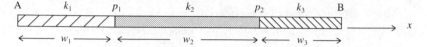

Figure 4.10   Horizontal cross-section of an insulating wall

$u^* = 50$ and at the outer side B heat is lost through convection, with $h = 50$ and $u_\infty = 26$. Expressed as a differential equation, u satisfies

$$\frac{d}{dx}\left(k\frac{du}{dx}\right) = 0 \quad 0 < x < 0.13$$

with boundary conditions $u(a) = u^*$, and

$$k\frac{du}{dx} \text{ is continuous at } P_1, P_2, \quad -k_3\frac{du}{dx}(B) = h(u(B) - u_\infty).$$

Alternatively, the problem may be expressed in variational form as: u satisfies $u(0) = u^*$ and minimises the functional

$$V((u)) = \int_0^{0.13} \frac{1}{2}ku'^2 \, dx + \frac{h}{2}u^2(B) - hu_\infty u(B).$$

(a) Obtain the exact solution by solving the differential equation.

(b) Calculate the three-element solution.

Compare the two results.

5 Two-dimensional heat flow

5.1 Introduction

This chapter extends the setting of the ideas which have been introduced previously, from one dimension into two. No new concepts are needed: functionals to be minimised, element approximation using shape functions, element stiffness matrices in condensed form and then expanded, and the global set of equations, all have been met before. They are met again here in a wider context.

The chapter ends with a flow chart showing the outline of the finite element algorithm. So the chapter:

- Applies linear triangular elements to solving Laplace's equation in the context of heat flow, in two dimensions.
- Solves a simple problem to illustrate the method.
- Considers essential and natural boundary conditions.

5.2 The model problem and an overview

Heat flow, rather than elasticity, has been chosen to illustrate the development into two dimensions because of its relative simplicity. In heat flow the problem variable is a scalar, temperature; in the case of elasticity, it is a vector quantity, displacement, requiring both x and y components in a two-dimensional analysis. This is considered in chapter 9.

As an illustrative problem, consider a long prism of rectangular cross-section, with a wire running through it parallel to its length, to provide a heat source. For a long prism, symmetry will allow the solution to be obtained from analysing the temperature distribution on a cross-section, thus reducing the problem to two dimensions. The units of the problem data may have to be modified by considering a unit thickness of the prism; for example a surface heat input, given as flux per area, is converted to flux per length of boundary. The boundary conditions are chosen to illustrate the main types that are commonly implemented. They are shown in figure 5.1. The solution of the problem requires finding the temperature distribution over the rectangle; once this has been obtained the heat flow may be calculated from it. A hypothetical (though it is hoped, not unreasonable) solution of the problem is shown in figure 5.2, where the temperature is displayed as the height of a smooth surface drawn above the cross-section. The finite element method tries to approximate to this surface.

This is done by first discretising the region (rectangle) into elements. The most commonly used shapes for elements in two dimensions are either triangles or

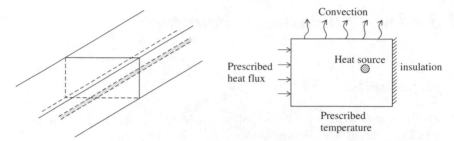

Figure 5.1   The model problem to illustrate the main ideas

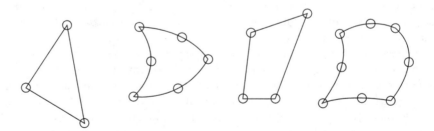

Figure 5.2   A hypothetical solution

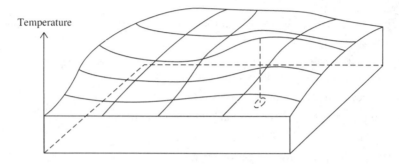

Figure 5.3   Common two-dimensional elements

quadrilaterals, with straight or curved sides (figure 5.3). The simplest is the straight-sided triangle which is defined by its three vertices. These are used as nodes, and are the positions at which the unknown temperature is to be estimated. The shapes of elements and the form of the approximation to the problem variable can be chosen in varying degrees of sophistication, but in this introduction to the subject straight-sided triangles will be mainly used. The approximation over each element will be linear, so that if the temperature is represented graphically it will form a

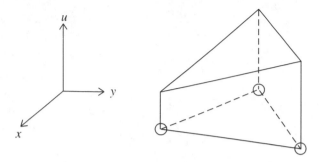

Figure 5.4 The approximation corresponding to a linear element

plane surface above the element with the height representing the temperature (figure 5.4). But more of this later in the section.

A simple mesh for the model problem is shown in figure 5.5, where care has been taken to site a node at the heat source. Suppose that the finite element solution has been obtained; what *kind* of solution will it be? Leaving aside the problem of accuracy (which is a difficult question to answer), it is important to understand the *nature* of the surface – will it be continuous and, if so, how smooth?

As shown in figure 5.4, over each element the solution surface will be flat and will be uniquely defined by the temperature values, u_i, u_j, u_k calculated at the vertices. For two adjacent elements the approximations will be combined as shown in figure 5.6 Since the temperatures at the common nodes i and j are the same for both elements, and the temperature surface above the line joining the nodes is a straight line joining points corresponding to the temperature values, it

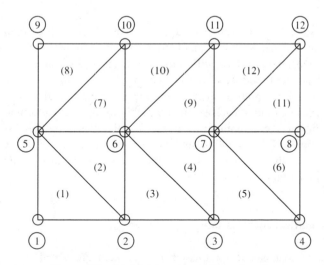

Figure 5.5 Discretising the problem domain

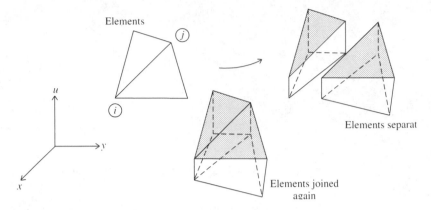

Figure 5.6   How element approximations combine

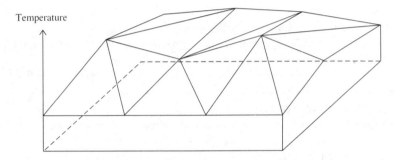

Figure 5.7   The form of a linear finite element prediction

will be the same line for both elements. Thus, the flat surfaces for each element will meet in a common line and blend into a continuous surface. Hence the complete temperature surface will consist of a set of flat triangular planes, and adjacent planes will meet in common lines as shown in figure 5.7.

To summarise: the finite element method using linear triangles gives rise to flat surfaces corresponding to each element which join together to form a continuous approximation to the smooth surface which is the exact solution. Note that the surface of figure 5.7 is continuous, but that its slope is not; there will be a discontinuity in the gradient at the element joins. The finite element solution obtained is continuous and piecewise linear.

5.3 Mathematical formulations

Two different mathematical forms of the problem are stated: (a) as a partial differential equation, and (b) in the variational form. Before returning to the problem

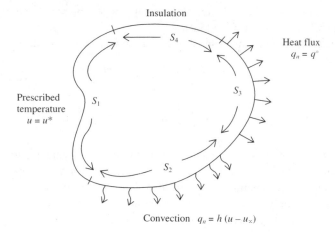

Figure 5.8   The problem domain with various boundary conditions

of section 5.2, the setting is generalised for a region R (figure 5.8). The two-dimensional aspect will arise from a three-dimensional problem with symmetry, resulting in the heat flow being modelled by a cross-section, or possibly the problem may be for a thin lamina insulated on its top and bottom surfaces.

Consider the temperature distribution $u(x, y)$ over the conducting region R with conductivity k. The resulting heat flow is $\mathbf{q} = -k\nabla u$, and in the direction normal to the boundary this becomes $q_n = -k(\partial u/\partial n)$ where $\mathbf{n}$ is the normal. There is a heat source $f(x, y)$ per unit area and a variety of possible boundary conditions as shown. The mathematical model will be stated both as a differential equation and in the variational form which is more suited to the finite element method.

(a) As a differential equation
The temperature $u(x, y)$ satisfies at all points in R, **Poisson's equation**

$$-k\left(\frac{\partial^2 u}{\partial x^2} + \frac{\partial^2 u}{\partial y^2}\right) = f(x, y) \tag{5.1}$$

and on the various parts of the boundary,

$S_1 : u = u^*$ u^* is the prescribed temperature
$S_2 : q_n = h(u - u_\infty)$ h is the convection coefficient and u_∞ is
                         the outside temperature
$S_3 : q_n = q^*$ q^* is the prescribed flux
$S_4 : q_n = 0$ an insulated boundary

(b) In variational form
The variational form is quoted at this stage but will be derived later in chapter 6. Of the different types of boundary conditions, the essential condition $u = u^*$ on

S_1 is 'built into' the solution and the others are imposed (approximately) by means of terms in the functional.

The temperature $u(x, y)$ satisfies $u = u^*$ on S_1 and minimises

$$V(u(x, y)) = \int \int_R \frac{k}{2}\left[\left(\frac{\partial u}{\partial x}\right)^2 + \left(\frac{\partial u}{\partial y}\right)^2\right] - uf(x, y)\,\mathrm{d}A$$

$$+ \int_{S_2} \frac{h}{2}(u - u_\infty)^2\,\mathrm{d}s + \int_{S_3} uq^*\,\mathrm{d}s \qquad (5.2)$$

Note that there is no boundary line integral on S_4 which corresponds to the insulated part of the boundary. This may be thought of as an applied flux of zero.

A modified problem is now considered, simplified by having no internal heat source and only essential boundary conditions. How the other features are incorporated into the calculation will be considered later.

5.4 The simplified problem

Suppose the dimensions of the rectangle are 3×2 and the prescribed boundary temperatures are as shown in figure 5.9. It is not clear how the two different boundary values $u = 100$ and $u = 0$ merge, so suppose that $u = 0$ applies along the right-hand edge and just round the two corners. A simple division into elements which reflects the symmetry and a numbering are shown in the earlier figure 5.5. Because of the symmetry about the centre line only half of the rectangle needs to be analysed. The computation involves elements 1–6 and nodes 1–8.

As a result of the simplified boundary conditions, the functional (5.2) becomes the double integral

$$V(u(x, y)) = \int \int_R \frac{k}{2}\left[\left(\frac{\partial u}{\partial x}\right)^2 + \left(\frac{\partial u}{\partial y}\right)^2\right]\,\mathrm{d}A \qquad (5.3)$$

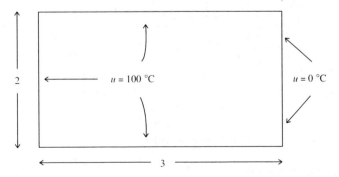

Figure 5.9   The simplified problem

By the property of integration, the integral over R may be obtained from the sum of integrals over subregions making up R

$$V = \int\!\!\int_{R^1} *\mathrm{d}A + \int\!\!\int_{R^2} *\mathrm{d}A + \cdots + \int\!\!\int_{R^e} *\mathrm{d}A + \cdots + \int\!\!\int_{R^6} *\mathrm{d}A$$

where the integrands are as in (5.3), and R^e is the region occupied by the element e. This may be expressed as

$$V = V^1 + V^2 + \cdots + V^e + \cdots + V^6 \tag{5.4}$$

where V^e is the functional ('energy') corresponding to the element e.

5.5 Calculations for a general triangular element

We need to obtain V^e and its derivatives for each element, which initially requires an integration of the form

$$V^e = \int\!\!\int_{R^e} \frac{k}{2}\left[\left(\frac{\partial u}{\partial x}\right)^2 + \left(\frac{\partial u}{\partial y}\right)^2\right]\mathrm{d}A \tag{5.5}$$

The section is separated into two parts; the first gives a general theory for linear triangular elements, and the second is a more earthy approach obtaining a simple matrix for the particular elements of the current problem.

5.5.1 A general linear element formulation

Consider the general element shown in figure 5.10 with nodes i, j and k. The temperature, u, is assumed to vary linearly on R^e

$$u^e(x, y) = \alpha x + \beta y + \gamma \tag{5.6}$$

where the constants α, β and γ are chosen to give the appropriate nodal values

$$u^e(\text{node } i) = u_i \qquad u^e(\text{node } j) = u_j \qquad u^e(\text{node } k) = u_k \tag{5.7}$$

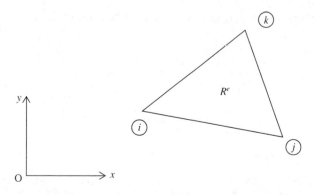

Figure 5.10   The general linear triangular element

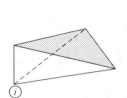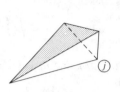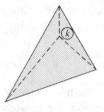

Figure 5.11   Linear shape functions

It is straightforward to calculate α, β and γ satisfying these conditions by setting up and solving three equations; however, the previous approach of section 3.4.1 using shape functions is preferred. The shape functions are chosen to be linear and to satisfy fundamental properties, which are, for $N_i(x, y)$ the shape function corresponding to node i, that

$$N_i(x, y) = \begin{cases} 1 & (x, y) \text{ at node } i \\ 0 & (x, y) \text{ at node } j \\ 0 & (x, y) \text{ at node } k \end{cases} \tag{5.8}$$

$N_j(x, y)$ and $N_k(x, y)$ satisfy corresponding properties. The properties have the effect of ensuring that the three shape functions (figure 5.11) are linearly independent and so can be used to construct a general linear expression. They also result in the simple form for the element approximation shown in equation (5.9).

It is not difficult to form the equations of the shape functions and the process will be explained in detail when the right-angled isosceles triangles are considered in section 5.5.2. It is left as an exercise to show that the linear form (5.6) satisfying (5.7) may be written as

$$u^e(x, y) = N_i(x, y)u_i + N_j(x, y)u_j + N_k(x, y)u_k \tag{5.9}$$

Exercise 5.1
Verify that u^e defined by (5.9) is (a) linear, and (b) satisfies (5.7).

From (5.9)

$$\frac{\partial u^e}{\partial x} = \frac{\partial N_i}{\partial x}u_i + \frac{\partial N_j}{\partial x}u_j + \frac{\partial N_k}{\partial x}u_k$$

$$\frac{\partial u^e}{\partial y} = \frac{\partial N_i}{\partial y}u_i + \frac{\partial N_j}{\partial y}u_j + \frac{\partial N_k}{\partial y}u_k$$

and substituting these into (5.5),

$$\frac{2}{k}V^e = \int\!\!\int_{R^e} \left[\left(\frac{\partial N_i}{\partial x}u_i + \frac{\partial N_j}{\partial x}u_j + \frac{\partial N_k}{\partial x}u_k \right)^2 + \left(\frac{\partial N_i}{\partial y}u_i + \frac{\partial N_j}{\partial y}u_j + \frac{\partial N_k}{\partial y}u_k \right)^2 \right] dA$$

and differentiating with respect to u_i in anticipation of the optimisation process,

$$\frac{1}{k}\frac{\partial V^e}{\partial u_i} = \int\int_{R^e}\left[\left(\frac{\partial N_i}{\partial x}u_i + \frac{\partial N_j}{\partial x}u_j + \frac{\partial N_k}{\partial x}u_k\right)\frac{\partial N_i}{\partial x}\right.$$

$$\left. + \left(\frac{\partial N_i}{\partial y}u_i + \frac{\partial N_j}{\partial y}u_j + \frac{\partial N_k}{\partial y}u_k\right)\frac{\partial N_i}{\partial y}\right]dA$$

or, briefly

$$\frac{\partial V^e}{\partial u_i} = K_{ii}^e u_i + K_{ij}^e u_j + K_{ik}^e u_k$$

where

$$K_{ii}^e = \int\int_{R^e} k\left[\frac{\partial N_i}{\partial x}\frac{\partial N_i}{\partial x} + \frac{\partial N_i}{\partial y}\frac{\partial N_i}{\partial y}\right]dA$$

$$K_{ij}^e = \int\int_{R^e} k\left[\frac{\partial N_j}{\partial x}\frac{\partial N_i}{\partial x} + \frac{\partial N_j}{\partial y}\frac{\partial N_i}{\partial y}\right]dA$$

$$K_{ik}^e = \int\int_{R^e} k\left[\frac{\partial N_k}{\partial x}\frac{\partial N_i}{\partial x} + \frac{\partial N_k}{\partial y}\frac{\partial N_i}{\partial y}\right]dA \qquad (5.10)$$

Exercise 5.2
Obtain the results corresponding to (5.10) for

$$\frac{\partial V^e}{\partial u_j} \quad \text{and} \quad \frac{\partial V^e}{\partial u_k}$$

Combining these results into matrix form

$$\begin{bmatrix} \frac{\partial V^e}{\partial u_i} \\ \frac{\partial V^e}{\partial u_j} \\ \frac{\partial V^e}{\partial u_k} \end{bmatrix} = \begin{bmatrix} K_{ii} & K_{ij} & K_{ik} \\ K_{ji} & K_{jj} & K_{jk} \\ K_{ki} & K_{kj} & K_{kk} \end{bmatrix} \begin{bmatrix} u_i \\ u_j \\ u_k \end{bmatrix}$$

or, briefly

$$\frac{\partial V^e}{\partial \mathbf{u}^e} = \mathbf{K}_c^e \mathbf{u}^e \qquad (5.11)$$

Note that $\mathbf{K}_c^e$ is symmetrical.

Here the 'c' indicates that the stiffness matrix is in condensed form. The general term of $\mathbf{K}_c^e$, K_{rc}, (the 'c' here denotes the column position, and 'r' the

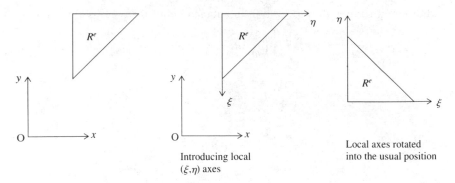

Local axes rotated
into the usual position

Introducing local
(ξ,η) axes

Figure 5.12   An element with more convenient axes

row) is given by

$$K_{rc}^e = \int\int_{R^e} k\left[\frac{\partial N_c}{\partial x}\frac{\partial N_r}{\partial x} + \frac{\partial N_c}{\partial y}\frac{\partial N_r}{\partial y}\right]dA \quad \text{where } r,c \text{ are } i,j \text{ or } k \qquad (5.12)$$

5.5.2 Computation for a particular 'general' element

In the discretisation of the model problem all the elements have the same geometry. They are all isosceles right-angled triangles with the same dimensions. In addition, the integrand in (5.5) is independent of the choice of axis. Consider

$$\left(\frac{\partial u}{\partial x}\right)^2 + \left(\frac{\partial u}{\partial y}\right)^2;$$

this is the squared magnitude of the vector $\nabla u = \frac{\partial u}{\partial x}\mathbf{i} + \frac{\partial u}{\partial y}\mathbf{j}$. Now ∇u is the greatest slope of the surface u at a point, an invariant quantity. This means that the axes may be changed in position and rotated without altering the integrand. New axes will be introduced in order to simplify the calculation (figure 5.12).

$$V^e = \int\int_{R^e} \frac{k}{2}\left[\left(\frac{\partial u}{\partial x}\right)^2 + \left(\frac{\partial u}{\partial y}\right)^2\right]dA = \int\int_{R^e} \frac{k}{2}\left[\left(\frac{\partial u}{\partial \xi}\right)^2 + \left(\frac{\partial u}{\partial \eta}\right)^2\right]dA$$

In the general finite element method with irregular element geometry, the calculations are made in a similar way on a general or 'master' element of simple shape. The results then have to be related back on to each element, using an appropriate mapping – see chapter 8.

5.5.3 Forming the approximation for $\mathbf{u}^e(\xi, \eta)$

Consider the isosceles right-angled element with equal sides of length l shown in figure 5.13.

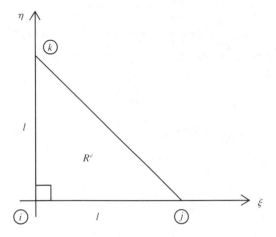

Figure 5.13 The right-angled isoceles element

The temperature, u^e, is assumed to vary linearly on R^e and may be constructed using shape functions formed with the (ξ, η) axes, by

$$u^e(\xi, \eta) = N_i(\xi, \eta)u_i + N_j(\xi, \eta)u_j + N_k(\xi, \eta)u_k$$

In order to carry out the calculation, we need to find expressions for the shape functions. Consider $N_i(\xi, \eta)$, the shape function corresponding to node i. This is chosen to be linear and to satisfy the defining relations (5.8) (figure 5.14). As $N_i = 0$ at nodes j and k, consider the equation of the line joining these nodes. The expression $\xi + \eta - l$ is linear and is zero at nodes j and k (and incidentally at all points on the line), so it satisfies two of the three requirements of (5.8). At node i, where $\xi = \eta = 0$, the expression $\xi + \eta - l$ becomes $-l$, and not 1 as required, but this can be adjusted by dividing through by $-l$, thus

$$N_i(\xi, \eta) = \frac{\xi + \eta - l}{-l} = \frac{1}{l}(l - \xi - \eta) \tag{5.13}$$

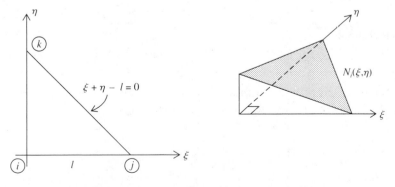

Figure 5.14 The linear shape function for node i

Exercise 5.3

(a) In a similar manner show that $N_j(\xi, \eta)$, which satisfies

$$N_j(\text{at node i}) = 0 \quad N_j(\text{at node j}) = 1 \quad N_j(\text{at node k}) = 0,$$

is given by
$$N_j(\xi, \eta) = \frac{\xi}{l} \tag{5.14}$$

(b) Write down the conditions to be satisfied by N_k and show that

$$N_k(\xi, \eta) = \frac{\eta}{l} \tag{5.15}$$

5.5.4 Calculating V^e

Using equations (5.13), (5.14) and (5.15) the derivatives that are needed for K_c are

$$\frac{\partial N_i}{\partial \xi} = -\frac{1}{l}, \quad \frac{\partial N_j}{\partial \xi} = \frac{1}{l}, \quad \frac{\partial N_k}{\partial \xi} = 0$$

$$\frac{\partial N_i}{\partial \eta} = -\frac{1}{l}, \quad \frac{\partial N_j}{\partial \eta} = 0, \quad \frac{\partial N_k}{\partial \eta} = \frac{1}{l}$$

Thus from (5.12)

$$K_{ii}^e = \int\int_{R^e} k \left[\frac{\partial N_i}{\partial x} \frac{\partial N_i}{\partial x} + \frac{\partial N_i}{\partial y} \frac{\partial N_i}{\partial y} \right] dA$$

$$= \int\int_{R^e} k \left[\left(-\frac{1}{l} \right)^2 + \left(-\frac{1}{l} \right)^2 \right] dA$$

$$= 2k \left(\frac{1}{l^2} \right) \int\int_{R^e} dA$$

$$= 2k \left(\frac{1}{l^2} \right) \left(\frac{l^2}{2} \right)$$

$$= k$$

Exercise 5.4

Show that $\quad K_{ij} = K_{ik} = -\frac{k}{2} \quad$ and $\quad K_{jk} = 0.$

The other terms in the stiffness matrix may be similarly obtained, so that the result (5.11) for the isosceles right-angled triangle is

$$\begin{bmatrix} \frac{\partial V^e}{\partial u_i} \\ \frac{\partial V^e}{\partial u_j} \\ \frac{\partial V^e}{\partial u_k} \end{bmatrix} = \frac{k}{2} \begin{bmatrix} 2 & -1 & -1 \\ -1 & 1 & 0 \\ -1 & 0 & 1 \end{bmatrix} \begin{bmatrix} u_i \\ u_j \\ u_k \end{bmatrix} \tag{5.16}$$

Exercise 5.5

Show that

$$\frac{\partial u^e}{\partial \xi} = \frac{1}{l}(-u_i + u_j)$$

$$\frac{\partial u^e}{\partial \eta} = \frac{1}{l}(-u_i + u_k)$$

Interpret these two results as finite differences. They show that, using linear elements, ∇u is assigned a constant value on R_e. If there is one position for it to be considered to act, where would that be?

5.6 The global finite element equations

Since V^e depends on its three nodal values, then for the whole region $V = \sum_{e=1}^{E} V^e$ will be formed of the values of u at all the nodes. That is,

$$V = V(u_1, u_2, \ldots, u_N),$$

where there are N nodes in the discretisation of the region. Thus, once the finite element approximation has been made, the search for a minimising function, u, (or surface) becomes a search for the minimising set of nodal values. So $u_1, u_2, \ldots, u_N$ are chosen to satisfy

$$\frac{\partial V}{\partial u_1} = \frac{\partial V}{\partial u_2} = \cdots = \frac{\partial V}{\partial u_N} = 0$$

Now

$$\frac{\partial V}{\partial u_1} = 0 \Rightarrow \frac{\partial V^1}{\partial u_1} + \frac{\partial V^2}{\partial u_1} + \cdots + \frac{\partial V^E}{\partial u_1} = 0$$

$$\frac{\partial V}{\partial u_2} = 0 \Rightarrow \frac{\partial V^1}{\partial u_2} + \frac{\partial V^2}{\partial u_2} + \cdots + \frac{\partial V^E}{\partial u_2} = 0$$

$$\vdots \qquad \vdots \quad \vdots \quad \ddots \quad \vdots \quad \vdots$$

$$\frac{\partial V}{\partial u_N} = 0 \Rightarrow \frac{\partial V^1}{\partial u_N} + \frac{\partial V^2}{\partial u_N} + \cdots + \frac{\partial V^E}{\partial u_N} = 0. \tag{5.17}$$

These are the set of N finite elements equations which, when solved with the appropriate boundary conditions, give the finite element solution to the problem.

5.7 The solution

As the simplified problem has symmetry about the centre line, only half of the region needs to be analysed.

5.7.1 The element stiffness matrices

The set of finite element equations (5.17), when written in the matrix notation introduced in section 4.2, become

$$\frac{\partial V^1}{\partial \mathbf{u}} + \frac{\partial V^2}{\partial \mathbf{u}} + \cdots + \frac{\partial V^6}{\partial \mathbf{u}} = \mathbf{0} \qquad (5.18)$$

where

$$\mathbf{u} = [u_1 \ u_2 \ \cdots \ u_8]^{\mathrm{T}} \text{ and } \frac{\partial V^e}{\partial \mathbf{u}} = \left[\frac{\partial V^e}{\partial u_1} \ \frac{\partial V^e}{\partial u_2} \ \cdots \ \frac{\partial V^e}{\partial u_8} \right]^{\mathrm{T}}$$

Writing the equations in this form emphasises that they may be constructed by adding the contribution of each element.

Consider the contribution of the first element, $\partial V^1 / \partial \mathbf{u}$. Comparing with the general element it will be seen that the general nodes i, j and k correspond to the particular nodes 1, 2 and 5 (figure 5.15).

Thus the result corresponding to equation (5.16) is

$$\begin{bmatrix} \frac{\partial V^1}{\partial u_1} \\ \frac{\partial V^1}{\partial u_2} \\ \frac{\partial V^1}{\partial u_5} \end{bmatrix} = \frac{k}{2} \begin{bmatrix} 2 & -1 & -1 \\ -1 & 1 & 0 \\ -1 & 0 & 1 \end{bmatrix} \begin{bmatrix} u_1 \\ u_2 \\ u_5 \end{bmatrix} \qquad (5.19)$$

Since V^1 is independent of the other nodal values u_3, u_4, u_6, u_7 and u_8

$$\frac{\partial V^1}{\partial u_3} = \frac{\partial V^1}{\partial u_4} = \cdots = \frac{\partial V^1}{\partial u_8} = 0 \qquad (5.20)$$

Combining (5.19) and (5.20) into matrix form gives the global equation

$$\frac{\partial V^1}{\partial \mathbf{u}} = \frac{k}{2} \begin{bmatrix} 2 & -1 & 0 & 0 & -1 & 0 & 0 & 0 \\ -1 & 1 & 0 & 0 & 0 & 0 & 0 & 0 \\ 0 & 0 & 0 & 0 & 0 & 0 & 0 & 0 \\ 0 & 0 & 0 & 0 & 0 & 0 & 0 & 0 \\ -1 & 0 & 0 & 0 & 1 & 0 & 0 & 0 \\ 0 & 0 & 0 & 0 & 0 & 0 & 0 & 0 \\ 0 & 0 & 0 & 0 & 0 & 0 & 0 & 0 \\ 0 & 0 & 0 & 0 & 0 & 0 & 0 & 0 \end{bmatrix} \begin{bmatrix} u_1 \\ u_2 \\ u_3 \\ u_4 \\ u_5 \\ u_6 \\ u_7 \\ u_8 \end{bmatrix}$$

$$= \mathbf{K}^1 \mathbf{u}$$

where $\mathbf{K}^1$ is called the stiffness matrix of element 1.

For the element 2 the relation with the general element is shown in figure 5.16. Using equation (5.16) the element stiffness matrices for element 2 may similarly be

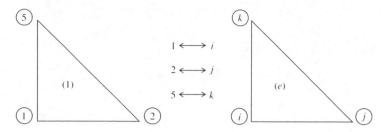

Figure 5.15 Relating element 1 with the general element

formed; this time i, j, k correspond to 6, 5, 2. Note that the order 6, 2, 5 will lead to the same stiffness matrix; it is the position of the 'right-angle' node which is significant. The details are left as an exercise.

Exercise 5.6
Obtain $\mathbf{K}^2$ and $\mathbf{K}^3$.

Continuing this process the other element stiffness matrices $\mathbf{K}^4$, $\mathbf{K}^5$ and $\mathbf{K}^6$ can be obtained. Return now to equation (5.17). Since

$$\frac{\partial V^e}{\partial \mathbf{u}} = \mathbf{K}^e \mathbf{u} \quad e = 1, 2, \ldots, 6 \qquad \text{and} \qquad \frac{\partial V}{\partial \mathbf{u}} = \mathbf{0}$$

then
$$\mathbf{K}^1 \mathbf{u} + \mathbf{K}^2 \mathbf{u} + \mathbf{K}^3 \mathbf{u} + \cdots + \mathbf{K}^6 \mathbf{u} = \mathbf{0}$$
$$\text{or} \quad \mathbf{K} \mathbf{u} = \mathbf{0} \qquad (5.21)$$

where $\mathbf{K} = \sum_{e=1}^{6} \mathbf{K}^e$ is the global stiffness matrix formed by adding together the element stiffness matrixes.

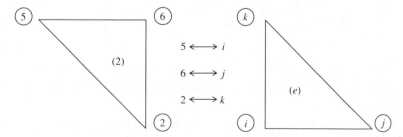

Figure 5.16 Relating element 2 with the general element

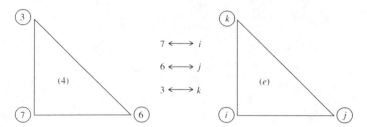

Figure 5.17   Rotating and relating element 4

5.7.2 Adding condensed K

Look again at equations (5.16)

$$
\begin{bmatrix}
\dfrac{\partial V^e}{\partial u_i} \\[2mm]
\dfrac{\partial V^e}{\partial u_j} \\[2mm]
\dfrac{\partial V^e}{\partial u_k}
\end{bmatrix}
= \frac{k}{2}
\begin{bmatrix}
2 & -1 & -1 \\
-1 & 1 & 0 \\
-1 & 0 & -1
\end{bmatrix}
\begin{bmatrix}
u_i \\
u_j \\
u_k
\end{bmatrix}
$$

$$
= \mathbf{K}_c^e \mathbf{u}^e
$$

Formally, the process of adding $\mathbf{K}^e$ into the global stiffness matrix, as described by equation (5.21), occurs after $\mathbf{K}_c^e$ has been expanded to the same dimensions as the global matrix by adding appropriate rows and columns of zeros. Clearly, the expansion is not necessary in practice; the condensed form may be added in directly, once the positions of the terms of $\mathbf{K}_c^e$ in the global structure have been determined.

Consider, for example, the contribution of element 4. The element is rotated to match e and put alongside in figure 5.17. Since $\partial V^e/\partial u_r$ gives rise to the rth row of the global matrix, and the cth column multiplies u_r, it follows that the positions for the terms of the condensed matrix for element 4 are

$$
\begin{array}{cccc}
 & \text{7th col} & \text{6th col} & \text{3rd col} \\
\begin{array}{c}\text{7th row} \\ \text{6th row} \\ \text{3rd row}\end{array} &
\left(\begin{array}{ccc}
2 & -1 & -1 \\
-1 & 1 & 0 \\
-1 & 0 & 1
\end{array}\right) & & \dfrac{k}{2}
\end{array}
$$

Exercise 5.7 is given in order to practice the technique of finding these positions. This is the method used by computer programs to construct the global stiffness matrix.

Exercise 5.7
Write out $\mathbf{K}_c^5$ and $\mathbf{K}_c^6$ indicating the row and column positions corresponding to the global matrix.

Exercise 5.8

By adding the element stiffness matrixes verify that the global equations are

$$
\begin{array}{llllllll}
2u_1 & -u_2 & & & -u_5 & & & & = 0 \\
-u_1 & 4u_2 & -u_3 & & & -2u_6 & & & = 0 \\
& -u_2 & u_3 & -u_4 & & & -2u_7 & & = 0 \\
& & -u_3 & 2u_4 & & & & -u_8 & = 0 \\
-u_1 & & & & 2u_5 & -u_6 & & & = 0 \\
& -2u_2 & & & -u_5 & 4u_6 & -u_7 & & = 0 \\
& & -2u_3 & & & -u_6 & 4u_7 & -u_8 & = 0 \\
& & & -u_4 & & & -u_7 & 2u_8 & = 0
\end{array}
\tag{5.22}
$$

Exercise 5.9

Given a condensed stiffness matrix, $\mathbf{K}_c^e$, consider how an algorithm could be written which would add the stiffness matrix of an element, t, of the same shape into the global matrix. Suppose a general element stiffness matrix is given by

$$
\mathbf{K}_c^e = \begin{bmatrix} K_{ii} & K_{ij} & K_{ik} \\ K_{ji} & K_{jj} & K_{jk} \\ K_{ki} & K_{kj} & K_{kk} \end{bmatrix}
$$

and t is a particular element (figure 5.18). Write down the increased values of $\mathbf{K}$ after the element stiffness matrix has been added, i.e. $K(7,7) = K(7,7) + K_{ii}$, etc.

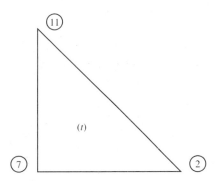

Figure 5.18   An element t

To complete the solution of the problem the boundary conditions must be applied and the resulting equations solved. The nodal values, u_1, u_2, u_3, u_4, u_5 and u_8 are given and are not free to vary, so the equations corresponding to differentiating with respect to these values are not relevant. Only the equations corresponding to $\partial V/\partial u_6 = 0$ and $\partial v/\partial u_7 = 0$ remain. That is,

$$
\begin{array}{llllllll}
-u_1 & -2u_2 & & & -u_5 & 4u_6 & -u_7 & & = 0 \\
& & -2u_3 & & & -u_6 & 4u_7 & -u_8 & = 0
\end{array}
$$

as $u_1 = u_2 = u_3 = u_5 = 100$ and $u_4 = u_8 = 0$, these become

$$4u_6 - u_7 = 300$$
$$-u_6 + 4u_7 = 200$$

which, when solved, give $u_6 = 93.3$ and $u_7 = 73.3$.

The simplified example considered has used only essential boundary conditions. Natural boundary conditions are incorporated through additional terms in the functional. How this happens will now be considered for a triangle of general shape.

5.8 Convection

The term in the functional which allows for convection (see section 5.3) is the line integral

$$V_c = \int_{S_2} \frac{h}{2} (u - u_\infty)^2 \, ds$$

Consider an element e with a side joining nodes i and j which is on (or approximately on) the boundary (figure 5.19). In this case, the above equation becomes

$$V_c^e = \int_{S^e} \frac{h}{2} (u^e - u_\infty)^2 \, ds$$

where $u^e = N_i u_i + N_j u_j + N_k u_k$ and N_i, N_j and N_k are the appropriate shape functions which were sketched in figure 5.11, and S^e is the straight line joining nodes i and j.

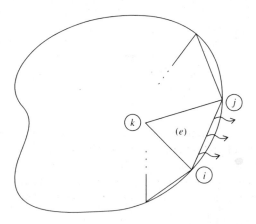

Figure 5.19   An element experiencing convection

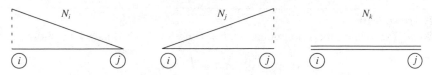

Figure 5.20 The shape functions corresponding to side i, j

So V_c^e becomes

$$\int_{S^e} \frac{h}{2}(N_i u_i + N_j u_j + N_k u_k - u_\infty)^2 ds$$

Then

$$\frac{\partial V_c^e}{\partial u_i} = \int_{S^e} h(N_i u_i + N_j u_j + N_k u_k - u_\infty)N_i \, ds$$

$$= \left[h \int_{S^e} N_i^2 \, ds \right] u_i + \left[h \int_{S^e} N_j N_i \, ds \right] u_j$$

$$+ \left[h \int_{S^e} N_k N_i \, ds \right] u_k - h u_\infty \int_{S^e} N_i \, ds \qquad (5.23)$$

The line integrals are not difficult if figure 5.15 is used to assess what the shape functions look like corresponding to the side joining nodes i and j. Suppose the side has length l. By considering the area under the graphs (figure 5.20)

$$\int_{S^e} N_i \, ds = \int_{S^e} N_j \, ds = \frac{l}{2}, \quad \int_{S^e} N_k \, ds = 0$$

These integrals will be needed later. To find the integrals in (5.23) it is helpful to express the other shape functions in terms of local coordinates (figure 5.21). On the side joining nodes i and j, $N_i = \frac{l-\xi}{l}$, so

$$\int_{S^e} N_i^2 \, ds = \int_0^l \left(\frac{l-\xi}{l} \right)^2 d\xi = \left[-\frac{1}{l^2} \frac{(l-\xi)^3}{3} \right]_0^l = \frac{l}{3}$$

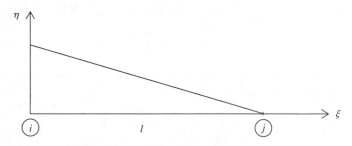

Figure 5.21 Local axes for N_i

Exercise 5.10

Write down N_j on the side joining i and j using local coordinates and show that

$$\int_{S^e} N_j^2 \, ds = \frac{l}{3} \quad \text{and} \quad \int_{S^e} N_i N_j \, ds = \frac{l}{6}$$

As $N_k = 0$ on S^e, all integrals involving N_k will be zero. Combining all these results, (5.23) gives

$$\frac{\partial V_c^e}{\partial u_i} = \frac{hl}{3} u_i + \frac{hl}{6} u_j - \frac{hl}{2} u_\infty$$

Exercise 5.11

Show that

$$\frac{\partial V_c^e}{\partial u_j} = \frac{hl}{6} u_i + \frac{hl}{3} u_j - \frac{hl}{2} u_\infty$$

and

$$\frac{\partial V_c^e}{\partial u_k} = 0$$

Hence convection along the side joining nodes i and j gives the additional finite element contribution

$$\frac{\partial V_c^e}{\partial \mathbf{u}^e} = \begin{bmatrix} \frac{hl}{3} & \frac{hl}{6} & 0 \\ \frac{hl}{6} & \frac{hl}{3} & 0 \\ 0 & 0 & 0 \end{bmatrix} \begin{bmatrix} u_i \\ u_j \\ u_k \end{bmatrix} - \begin{bmatrix} \frac{hl}{2} u_\infty \\ \frac{hl}{2} u_\infty \\ 0 \end{bmatrix} \tag{5.31}$$

The matrix is added to the element stiffness matrix and the vector to the force vector.

5.9 Flux on the boundary

Suppose that there is a known flux *input* on the boundary which is assumed constant, $-q^*$ per unit length. Then the term in the functional allowing for this is

$$V_f = -\int_{S_3} u q^* \, ds$$

(Note the change of sign from (5.2) where the flux is leaving the domain.) For a general element similar to figure 5.19

$$V_f^e = -q^* \int_{S^e} u^e \, ds$$

The working, which proceeds as in (5.8), is left as an exercise.

Exercise 5.12

Using the shape functions of section 5.7, show that

$$\frac{\partial V_f^e}{\partial \mathbf{u}^e} = -\begin{bmatrix} \frac{l}{2}q^* \\ \frac{l}{2}q^* \\ 0 \end{bmatrix}$$

thus giving a contribution to the force vector.

5.10 A point source of heat within the body

In the model problem of section 5.2, there was a heat source within the body of the prism and care was taken to position a node at this point, say node i. The heat source is treated as a point source of heat of value f. The technique for modelling a point source is simple: the value f is added to the right-hand side of the ith equation. To explain the background theory requires the use of the Dirac impulse, and is omitted from this introductory book. If the point source is not at a node, then it is shared between the three nodes of the element in which it occurs, but how this is done is again beyond the scope of this book.

5.11 Summary

For each element, allowing for the possibility of convection, or flux input, the general form is

$$\frac{\partial V^e}{\partial \mathbf{u}} = \mathbf{K}^e \mathbf{u} - \mathbf{f}^e$$

where $\mathbf{K}^e$ is the element stiffness matrix and $\mathbf{f}^e$ is the element force vector. The finite element equations are obtained by summing the matrices.

Since

$$V = \sum_{e=1}^{E} V^e$$

then

$$\frac{\partial V}{\partial \mathbf{u}} = \sum_{e=1}^{E} \frac{\partial V^e}{\partial \mathbf{u}}$$

$$= \sum_{e=1}^{E} (\mathbf{K}^e \mathbf{u} - \mathbf{f}^e)$$

$$= \left[\sum_{e=1}^{E} \mathbf{K}^e \right] \mathbf{u} - \left[\sum_{e=1}^{E} \mathbf{f}^e \right]$$

$$= \mathbf{Ku} - \mathbf{f}$$

And since $\dfrac{\partial V}{\partial \mathbf{u}} = 0$ then $\mathbf{Ku} = \mathbf{f}$

where $\mathbf{K}$ is the global stiffness matrix, $\mathbf{f}$ is the global force vector and $\mathbf{u}$ is the vector of unknown nodal temperatures.

A flow sequence outlining the steps of a computer implementation of the finite element method is as follows:

- Read in structure data, element numbers and position of nodes, also material parameters.
- Do for each element:
  - calculate element stiffness matrix and force vector (condensed),
  - add into the appropriate positions of the global stiffness matrix and force vector.
- Continue.
- Modify global matrix and force vector to include the boundary conditions.
- Solve the resulting set of linear equations.
- Output results.

5.12 The model problem with numerical values

To draw together the various items of the analysis let us (mainly the reader!) work through the model problem using material parameters and boundary conditions, as shown in figure 5.22.

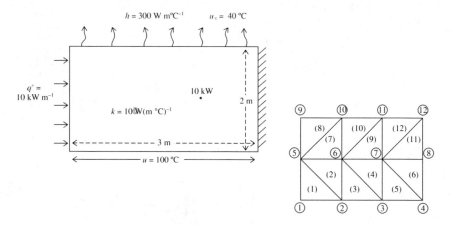

Figure 5.22   The model problem with specific boundary conditions

Exercise 5.13

Recalling previous results, show that where no convection is present, $\mathbf{K}_c^e$ is given by

$$\begin{bmatrix} 100 & -50 & -50 \\ -50 & 50 & 0 \\ -50 & 0 & 50 \end{bmatrix} \begin{bmatrix} u_i \\ u_j \\ u_k \end{bmatrix}$$

Also show the contribution from convection along the side joining nodes i and j is

$$\begin{bmatrix} 100 & 50 & 0 \\ 50 & 100 & 0 \\ 0 & 0 & 0 \end{bmatrix}$$

to be added to the stiffness matrix, and

$$\begin{bmatrix} 6000 \\ 6000 \\ 0 \end{bmatrix}$$

to be added to the force vector.

If the flux of 10 kW m^{-1} is entering the side joining nodes i and j normally, obtain the corresponding force vector,

$$\begin{bmatrix} 5000 \\ 5000 \\ 0 \end{bmatrix}$$

Exercise 5.14

Deduce the following finite element stiffness matrix and force vector. A factor of 50 has been taken out of the matrix and 1000 from the force vector. The essential boundary conditions have yet to be 'built in.'

$$\begin{bmatrix} 2 & -1 & 0 & 0 & -1 & 0 & 0 & 0 & 0 & 0 & 0 & 0 \\ -1 & 4 & -1 & 0 & 0 & -2 & 0 & 0 & 0 & 0 & 0 & 0 \\ 0 & -1 & 4 & -1 & 0 & 0 & -2 & 0 & 0 & 0 & 0 & 0 \\ 0 & 0 & -1 & 2 & 0 & 0 & 0 & -1 & 0 & 0 & 0 & 0 \\ -1 & 0 & 0 & 0 & 4 & -2 & 0 & 0 & -1 & 0 & 0 & 0 \\ 0 & -2 & 0 & 0 & -2 & 8 & -2 & 0 & 0 & -2 & 0 & 0 \\ 0 & 0 & -2 & 0 & 0 & -2 & 8 & -2 & 0 & 0 & -2 & 0 \\ 0 & 0 & 0 & -1 & 0 & 0 & -2 & 4 & 0 & 0 & 0 & -1 \\ 0 & 0 & 0 & 0 & -1 & 0 & 0 & 0 & 2 & -1 & 0 & 0 \\ 0 & 0 & 0 & 0 & 0 & -2 & 0 & 0 & 0 & 8 & 0 & 0 \\ 0 & 0 & 0 & 0 & 0 & 0 & -2 & 0 & 0 & 0 & 8 & 0 \\ 0 & 0 & 0 & 0 & 0 & 0 & 0 & -1 & 0 & 0 & 0 & 4 \end{bmatrix} \begin{bmatrix} 5 \\ 0 \\ 0 \\ 0 \\ 10 \\ 0 \\ 10 \\ 0 \\ 11 \\ 12 \\ 12 \\ 6 \end{bmatrix}$$

The essential boundary conditions are $u_1 = u_2 = u_3 = u_4 = 100$ (it may be disputed whether it is better to set $u_4 = 100$ or to leave it free to vary; either way a refinement of the mesh would tend towards a common result). After the equations have been suitably adjusted, the coefficient matrix and force vector are, with the same factors removed,

$$
\begin{bmatrix}
4 & -2 & 0 & 0 & -1 & 0 & 0 & 0 \\
-2 & 8 & -2 & 0 & 0 & -2 & 0 & 0 \\
0 & -2 & 8 & -2 & 0 & 0 & -2 & 0 \\
0 & 0 & -2 & 4 & 0 & 0 & 0 & -1 \\
-1 & 0 & 0 & 0 & 2 & -1 & 0 & 0 \\
0 & -2 & 0 & 0 & 0 & 8 & 0 & 0 \\
0 & 0 & -2 & 0 & 0 & 0 & 8 & 0 \\
0 & 0 & 0 & -1 & 0 & 0 & 0 & 4
\end{bmatrix}
\begin{bmatrix}
15 \\ 10 \\ 20 \\ 5 \\ 11 \\ 12 \\ 12 \\ 6
\end{bmatrix}
$$

The temperatures at the nodes are shown in figure 5.23. Once the temperatures have been calculated, further information can be derived and displayed. Finite element software usually performs this **post-processing** as part of the presentation of the solution; for heat flow, the isotherms would be displayed, and vectors showing the movement of heat, which is calculated from $\mathbf{q} = -k\nabla u$. Also, the amount of heat (input or output) which is needed to maintain a prescribed temperature value can be found. This is determined by the equations discarded at the time when the essential boundary conditions were being applied.

For example, consider node 1 where the temperature is 100 °C. Suppose that the heat input at the node required for this is q_1. Then the first equation is

$$100u_1 - 50u_2 - 50u_5 = 10\,000 + q_1$$

Substituting $u_1 = 100$, $u_2 = 100$, $u_5 = 151.0$ gives $q_1 = -7550$. This means that 7.55 kW of heat must be taken from the node in order that the temperature remains at 100 °C.

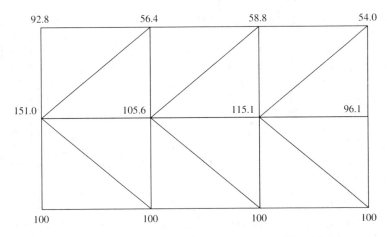

Figure 5.23   The solution temperatures at the nodes

General exercises for chapter 5

1. Explain how an insulated boundary is modelled, and what effect it has on the element equations.

2. The rectangular conducting sheet shown in figure 5.24 has dimensions 2×1. In the element discretisation shown, all triangles are right-angled and isosceles. Write down the finite element equations for the temperatures at nodes 1, 2, 3 and 4, and solve them.

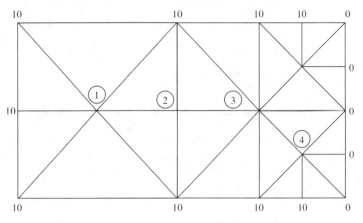

Figure 5.24   Exercise 2

3. Form the linear shape functions N_i, N_j and N_k for the equilateral triangle shown in figure 5.25.

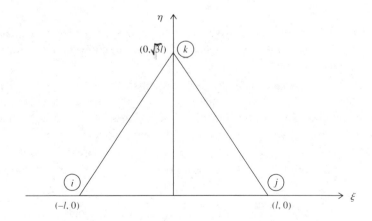

Figure 5.25   An equilateral triangular element

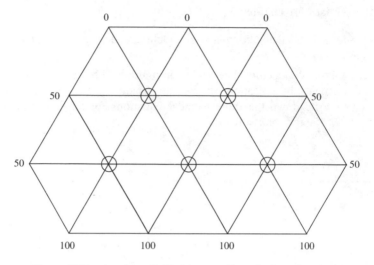

Figure 5.26   A region divided into equilateral triangular elements

4. Form the stiffness matrix for the three-noded element of exercise 3 by minimising the functional

$$\int\int_{R^e} \frac{k}{2}\left[\left(\frac{\partial u}{\partial \xi}\right)^2 + \left(\frac{\partial u}{\partial \eta}\right)^2\right] dA$$

5. Find the finite element approximation to the temperatures at nodes marked ○ in figure 5.26. All the elements are equilateral triangles of side length 1, and the boundary values are as shown.

6. The regular hexagon with sides of unit length shown in figure 5.27 is made of material of thermal conductivity $k = 150$. There is a heat flux input $q^* = 15\,000/\sqrt{3}$ per unit length on side AB, and side CD is maintained at $u = 50$; the other sides are insulated. Obtain $u_1, u_2, \ldots, u_5$.

7. Show that the heat lost due to convection $\int_S q_n \, ds$ is modelled for the side of a linear element of length l joining nodes i and j, by

$$\text{Heat lost} = \int_{S^e} q_n \, ds = \frac{hl}{2}(u_i + u_j - 2u_\infty)$$

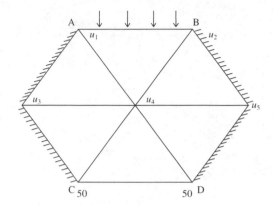

Figure 5.27   A regular haxagon

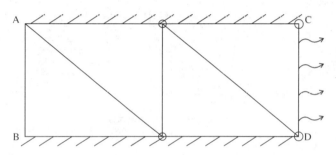

Figure 5.28   Exercise 8

8. The rectangle with sides AB of length 2 and AC = 4 shown in figure 5.28 is made of material of thermal conductivity $k = 100$. The side AB is maintained at temperature $u = 100$ and the side CD experiences convection, $h = 300$, $u_\infty = 60$, and the others are insulated. Use this with the model problem of section 5.2 to find the heat lost through convection. Find also the heat transferred at the nodes 2, 3 and 4. Verify the conservation of heat for the problem. Solve for the temperature at the ringed nodes. How much heat must be input along the side AB in order to maintain the temperature at A and B at 100 °C? Verify that heat is conserved in the finite element model.

9. Form the eight heat-flow equations for $u_1, u_2, \cdots, u_8$ at the nodes shown in figure 5.29. Find the solution (you will need software to solve the equations) and mark the temperatures against the nodes. Sketch the isotherms and show the heat flux through drawing a vector at each element centroid.

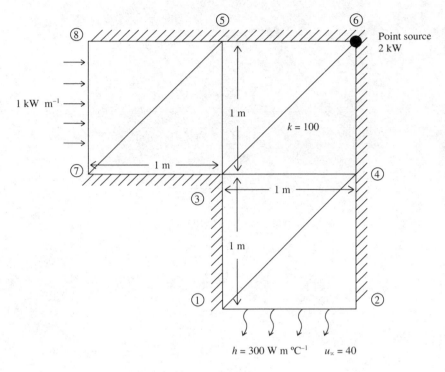

Figure 5.29   Exercise 9

6 Variational form

6.1 Introduction

The theme of this chapter is the method of changing the mathematical model of a problem from a differential equation into a variational form. The ability to make the change is probably not a skill which will be often called upon, but the results which have previously had to be taken on trust will be justified. In addition, an understanding of the process will give a deeper appreciation of the finite element method.

The process leads to a more generally applicable method of deriving the finite element equations than minimising a functional, since there are problems for which such a functional does not exist. The main ideas introduced are:

- The process of changing from a differential equation into a general variational form.

- How the simple variational form may often be converted into a form that is symmetrical in the trial and test functions.

- In some significant cases the weak variational form can be converted into the process of minimising a functional. This was the form that was used in previous chapters.

6.2 Differential equation and variational form

The natural setting for the mathematical formulation of the finite element method is for the problem to be stated in terms of minimising a functional (often a measure of energy). However, in many situations the modelling of the physical processes involved is expressed most naturally by means of differential equations. Physical laws are often statements about rates of change rather than about energy principles.

We shall illustrate with the two basic laws of heat transfer, where differential equations express the physical principles involved and there is no natural corresponding energy statement.

The law that 'heat flow is caused by a difference in temperature and is directly proportional to the difference' becomes, in mathematical language

$$\mathbf{q} = -k\nabla u$$

And the law that 'heat is conserved' becomes, if no source is present,

$$\nabla \cdot \mathbf{q} = 0$$

[where, in more detail, u is temperature; $\mathbf{q}$ the heat flow vector; ∇u the *gradient vector* $\dfrac{\partial u}{\partial x}\mathbf{i} + \dfrac{\partial u}{\partial y}\mathbf{j}$; $\nabla \cdot \mathbf{q}$ is the *divergence* of $\mathbf{q}$, $\left(\dfrac{\partial}{\partial x}\mathbf{i} + \dfrac{\partial}{\partial y}\mathbf{j}\right) \cdot (q_x\mathbf{i} + q_y\mathbf{j}) = \dfrac{\partial q_x}{\partial x} + \dfrac{\partial q_x}{\partial y}$].

In this way the problem is naturally described by differential equations, whereas if we wish to apply the finite element method the variational form is appropriate. This chapter describes the process of converting the mathematical model of a problem from one form to the other. The next section introduces some new terminology and a different way of looking at the process of obtaining a solution of a differential equation.

6.3 Trial functions and residuals

There are various analytical techniques for solving differential equations, which usually differ according to the form of the equation, for example,

$$u' - au = 0$$

can be solved by assuming that the solution has the form Ae^{mx}. Or, as another example,

$$u' + f(x)u = g(x)$$

can be solved by multiplying by an integrating factor $\exp(\int f(x)\,dx)$. Suppose a simple, more primitive approach is adopted, which does not assume any knowledge of these techniques. Make a sensible guess at the solution and then test to see how good it has been.

To illustrate, consider the boundary value problem

$$u'' + 3u = x \qquad 0 < x < 1 \qquad \text{where} \quad u(0) = u(1) = 0 \qquad (6.1)$$

Since the boundary values are known it is sensible to 'build' them into the **trial function**. Suppose we try the simple quadratic $g(x) = x(1 - x)$, which satisfies the given boundary conditions on u. Substituting into the differential equation, the left-hand side becomes

$$-2 + 3x(1 - x) = -3x^2 + 3x - 2$$

For the trial to be the solution, both sides of the equation must be the same, which means that the expression

$$-3x^2 + 3x - 2 \quad \text{should be} \quad x$$

Clearly it is not, and the difference between the two functions is called the **residual** $r(x)$ (figure 6.1).

In this case $r(x) = -3x^2 + 3x - 2 - x = -3x^2 + 2x - 2$.

If the guess by chance happened to be the solution, then the residual would be zero for all x in the interval [0,1]. The residual can be used as a measure of the

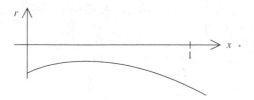

Figure 6.1   The residual

accuracy of the guess. It is not the **error**, which is defined as the difference between the exact and approximate solution, i.e.

$$e(x) = u(x) - g(x).$$

The residual and the error function are related, since if $g(x)$ is exact then both $r(x)$ and $e(x)$ are zero, but not in a simple manner. The residual $r(x)$ can be found, whereas generally the exact solution $u(x)$ and hence $e(x)$ cannot.

To return to the problem: if a constant factor α is introduced into the guess in an attempt to make it more flexible, so that it becomes $\alpha x(1 - x)$, we can choose α so that the residual is minimised in some way. There are several ways of doing this and the techniques are collectively called **weighted residual methods**.

The strength of the finite element method is that it generates a suitable set of trial functions defined by nodal values as parameters, together with a technique for choosing the best set of parameters.

Exercise 6.1
For the boundary value problem
$$u'' - u' + u = x^2 - 2x + 2 \qquad 0 < x < 1 \qquad \text{where } u(0) = 0, \quad u(1) = 1$$

form the residuals for the trial functions

(a) $g(x) = x$,   (b) $g(x) = x^2$,

and sketch their graphs.

Exercise 6.2
Make the trial $\alpha x(1 - x)$ in equation (6.1), determine the residual and calculate the value of α which makes $r(x) = 0$ at $x = 0.5$. Sketch the graph of the residual for this value of α.

Generalisation and summary
Use is made of the 'operator' notation, by which the symbol L is used in order to write a general differential equation as $L(u(x)) = f(x)$. For example $L = \mathrm{d}^2/\mathrm{d}x^2 + 3$ in equation (6.1)

Consider the differential equation $L(u(x)) = f(x) \qquad a < x < b$
with boundary conditions $u(a) = u_a, u(b) = u_b$.

A trial $u = g(x)$ is chosen which satisfies the boundary conditions,
so that $g(a) = u_a$ and $g(b) = u_b$.
The residual is then defined as $r(x) = L(g(x)) - f(x)$.
If $r(x) = 0$ on $[a, b]$ then $g(x)$ is a solution.

6.4 The fundamental lemma of the variational calculus

This lies at the heart of the change into variational form. In its simplest form it
states that if a function h is continuous, and if

$$\int_a^b h(x)v(x)\,dx = 0 \qquad \text{for all continuous functions } v,$$

then $h(x) \equiv 0$ for $a \le x \le b$. The statement that the integral should be zero *for all*
$v(x)$ is very demanding. It means that v can be chosen in a variety of ways and in
particular to highlight sections of the interval $[a, b]$ and 'force' $h(x)$ to be zero
there. The simplest proof, with the assumption that h is continuous, is to choose v
to be h. Then the integral becomes

$$\int_a^b h^2(x)\,dx = 0$$

and since $h^2(x) \ge 0$ for all x then there cannot be a point at which it is not zero,
so that $h(x) = 0$ for all points in the interval (this can be made into a rigorous
proof by using the definition of continuity). It should be mentioned that although
the theorem is stated with the assumption that $h(x)$ is continuous, we need to use
the theorem in circumstances where h has less continuity. In the finite element
method the trial functions are commonly piecewise linear or quadratic, and hence
the derivative is likely to have discontinuities at the element joins. Thus
$h(x) = L(g(x)) - f(x)$ may not qualify for the theorem as stated, but the
conditions on h can be suitably weakened.

The $h(x)$ of the theorem statement is chosen as the residual $r(x)$ which we wish to
be zero at all points of the problem domain, thereby forcing the trial function to be the
exact solution. The v are called the **weighting** or **test** functions.

6.5 Changing into a variational form

6.5.1 The one-dimensional case

Consider again the boundary value problem: to find a function u such that

$$L(u(x)) = f(x) \qquad \text{for } a < x < b$$

where $u(a) = u_a, u(b) = u_b$. As previously discussed in section 6.3, a trial
function $u(x) = g(x)$ is made which satisfies the boundary conditions. Then the

residual $r(x) = L(g(x)) - f(x)$ is formed. The idea of section 6.4 is now used in order to produce a condition to force $g(x)$ to satisfy the differential equation, i.e. so that $r(x) = 0$ on $a \le x \le b$. By the fundamental lemma, the trial function will be the solution if

$$\int_a^b r(x)v(x)\,dx = 0 \qquad \text{for all continuous } v \qquad (6.2)$$

Since the solution is known at both $x = a$ and $x = b$, it is usual and convenient to choose $v(x)$ to be zero at these two points. The trial is correct there already, and so does not need any forcing.

The statement (6.2) is the problem changed to **variational form**, though it will go through a further transformation before it is most useful.

Example: express in variational form the boundary value problem

$$u''(x) + au(x) = -f(x) \qquad 0 < x < 1 \qquad \text{where } u(0) = u(1) = 0$$

A trial function $u(x)$ is introduced. (We will use the same symbol 'u' instead of introducing a new name 'g' as in the previous paragraph. The notation is a little more elegant if not so precise.) The residual is now

$$r(x) = u''(x) + au(x) + f(x)$$

The variation form is: to find a trial function $u(x)$ satisfying $u(0) = 0$, $u(1) = 0$ and such that

$$\int_0^1 (u''(x) + au(x) + f(x))v(x)\,dx = 0$$

$$\text{for all continuous } v \text{ such that } v(0) = v(1) = 0 \qquad (6.3)$$

It might be added at this stage that the problem, although changed in form, is not any easier to solve!

Exercise 6.3

Express the following boundary value problems in variational form

(a) $AEu'' = -w$ $0 < x < 1$ where $u(0) = 0$, $u(1) = 1$.

(b) $uu'' + 3u' = 0$ $0 < x < 1$ where $u(0) = 0$, $u(1) = 1$.

(This is a non-linear equation because of the term uu''.)

6.5.2 Extension to two dimensions

To extend the ideas to two dimensions, consider the boundary value problem used in chapter 5, describing heat flow (figure 6.2)

$$\nabla^2 u = \frac{\partial^2 u}{\partial x^2} + \frac{\partial^2 u}{\partial y^2} = -f(x, y) \quad \text{for } (x, y) \text{ in } R$$

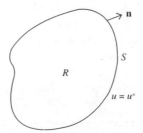

Figure 6.2   The domain for the boundary value problem

The essential boundary condition is that $u = u^*$ on S, the boundary of R. The development of a variational form carries over quite naturally from one dimension to two. The fundamental lemma becomes: given that a function H is continuous, and if

$$\int \int_R H(x, y)v(x, y)\,dA = 0$$

for all continuous functions v on R, then $H(x, y) = 0$ for (x, y) on R. This result will be assumed without further discussion.

To produce the variational form we proceed as before. Choose a trial function $u(x, y)$ satisfying $u(x, y) = u^*$ on S, and form the residual

$$r(x, y) = \nabla^2 u(x, y) + f(x, y) \tag{6.4}$$

The residual will be zero on R, and consequently the trial will be a solution, if

$$\int \int_R (\nabla^2 u(x, y) + f(x, y))v(x, y)\,dA = 0 \tag{6.5}$$

for all continuous functions v, where $v = 0$ on S, since u is known there.

6.6 Symmetry for trial and test functions

The variational form produced in the previous sections can be turned into a form which is more evenly balanced between the trial and test functions u and v. In the form given by (6.3) or (6.5) the residuals contain second derivatives of u, whereas the test function is unmodified; so a different smoothness (differentiability) is being required of the two functions. The process that provides a better balance is 'integration by parts' in one dimension, and its equivalent in two dimensions. It is used to transfer one of the derivatives of u onto v. In several cases of practical importance this produces a variational form which is symmetric in u and v.

As a preliminary, consider the process of integration by parts applied to a product of two functions, one of which is a second derivative.

In one dimension, integration by parts is

$$\int_a^b u'' v \, dx = [u'v]_a^b - \int_a^b u'v' \, dx. \tag{6.6}$$

In two or three dimensions there is a similar result

$$\iint_R (\nabla^2 u) v \, dA = \int_S \frac{\partial u}{\partial n} v \, ds - \iint_R (\nabla u) \cdot (\nabla v) \, dA \tag{6.7}$$

where the direction of **n** is the outward normal to the boundary S. This can be proved using a standard vector identity and the divergence theorem; the proof of the identity,

$$\nabla \cdot (v \nabla u) = \nabla v \cdot \nabla u + v \nabla^2 u \tag{6.8}$$

is required by exercise 6.4.

Exercise 6.4
Set the two-dimensional Cartesian form of ∇ in the left-hand side of (6.8) to show that

$$\nabla \cdot (v \nabla u) = \frac{\partial v}{\partial x} \frac{\partial u}{\partial x} + \frac{\partial v}{\partial x} \frac{\partial u}{\partial x} + v \frac{\partial^2 u}{\partial x^2} + v \frac{\partial^2 u}{\partial y^2}$$

and compare with the expanded form of the right-hand side.

The divergence theorem is probably well known:
For a suitably differentiable vector function $\mathbf{a}(x, y)$

$$\iint_R \nabla \cdot \mathbf{a} \, dA = \int_S \mathbf{a} \cdot \mathbf{n} \, ds \tag{6.9}$$

where **n** is the unit boundary normal. Now choosing $v \nabla u$ as the vector function **a**, (6.9) becomes

$$\iint_R \nabla \cdot (v \nabla u) \, dA = \int_S (v \nabla u) \cdot \mathbf{n} \, ds$$

and using the identity (6.8) in the left-hand side,

$$\iint_R (\nabla v \cdot \nabla u + v \nabla^2 u) \, dA = \int_S v (\nabla u \cdot \mathbf{n}) \, ds$$

Exercise 6.5
Write out in Cartesian form $\nabla v . \nabla u$ using

$$\nabla = \frac{\partial}{\partial x} \mathbf{i} + \frac{\partial}{\partial y} \mathbf{j}$$

Since $\nabla u \cdot \mathbf{n} = \partial u / \partial n$,

$$\int \int_R v \nabla^2 u \, dA = \int_S v \frac{\partial u}{\partial n} \, ds - \int \int_R (\nabla v \cdot \nabla u) \, dA \qquad (6.10)$$

We can now proceed to modify the variational forms given previously as (6.3) and (6.5).

6.6.1 The one-dimensional case with essential boundary conditions

As an example consider the boundary value problem

$$u''(x) + au(x) = -f(x) \qquad 0 < x < 1 \quad \text{where} \quad u(0) = u(1) = 0$$

The variational form is (see section 6.5): find $u(x)$ such that $u(0) = 0, u(1) = 0$ and

$$\int_0^1 (u''(x) + au(x) + f(x))v(x) \, dx = 0$$

$$\text{for all continuous } v \text{ such that } v(0) = v(1) = 0 \qquad (6.11)$$

Consider the term $\int_0^1 u''v \, dx$, using the integration by parts formula (6.6)

$$\int_0^1 u''v \, dx = [u'v]_0^1 - \int_0^1 u'v' \, dx$$

$$= u'(1)v(1) - u'(0)v(0) - \int_0^1 u'v' \, dx$$

$$= -\int_0^1 u'v' \, dx \qquad \text{since } v(0) = v(1) = 0 \qquad (6.12)$$

Thus (6.11) becomes

$$\int_0^1 (-u'v' + auv) \, dx = -\int_0^1 fv \, dx \qquad (6.13)$$

The form of the left-hand side of (6.11) has been changed in that one of the derivatives of u has been transferred to v; instead of the term $u''v$ we now have the symmetric term $u'v'$. It is also significant that the differentiability implicitly required by the differential equation, i.e. that u should be twice differentiable, has now been weakened in that only the first derivative appears. The form (6.13) is called the **weak variational form.**

6.6.2 The two-dimensional case with essential boundary conditions

The development is similar to that for one dimension and is left as an exercise.

Exercise 6.6
Show that the weak variational form corresponding to the boundary value problem (6.5) is: find u such that $u = u^*$ on S and

$$\int\int_R \nabla u \cdot \nabla v \, dA = -\int\int_R fv \, dA$$

for all v such that $v = 0$ on S.

6.6.3 Natural boundary conditions

The new feature introduced in this section is the addition of boundary integrals to the functional. This arises because the value of u is not now known explicitly at all points of the boundary; on part at least it is the derivative which is given. This has the implication that the test function v cannot be chosen to be zero on all of the boundary as it was before. The two-dimensional case is used to illustrate the ideas and the reader is left to work through a one-dimensional problem. Consider the boundary value problem (figure 6.3)

$$-k\nabla^2 u(x, y) = f(x, y) \qquad \text{for } (x, y) \text{ in } R$$

where

$$u = u^* \text{ on } S_1$$

$$-k\frac{\partial u}{\partial n} = q_n = q^* \text{ on } S_2$$

$$-k\frac{\partial u}{\partial n} = q_n = h(u - u_\infty) \text{ on } S_3$$

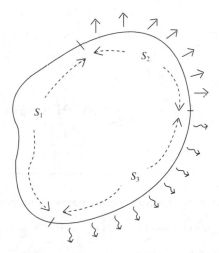

Figure 6.3   Various boundary conditions

The outward normal to the boundary of R, S, is $\mathbf{n}$ and $S_1 \cup S_2 \cup S_3 = S$. Continuing the previous approach, we build into the trial functions the values of u that are known; so u is chosen to satisfy $u = u^*$ on S_1 and

$$\int\int_R (-k\nabla^2 u - f)v \, dA = 0 \qquad \text{for all continuous } v \qquad (6.14)$$

where $v = 0$ on S_1.

Applying integration by parts (6.14) becomes

$$-\int_S kv \frac{\partial u}{\partial n} \, ds + \int\int_R (k\nabla v \cdot \nabla u) \, dA = \int\int_R fv \, dA$$

Because $S_1 \cup S_2 \cup S_3 = S$ it follows that

$$-\int_S kv \frac{\partial u}{\partial n} \, ds = -\int_{S_1} kv \frac{\partial u}{\partial n} \, ds - \int_{S_2} kv \frac{\partial u}{\partial n} \, ds - \int_{S_3} kv \frac{\partial u}{\partial n} \, ds$$

$$= 0 - \int_{S_2} kv \frac{\partial u}{\partial n} \, ds - \int_{S_3} kv \frac{\partial u}{\partial n} \, ds,$$

since the test function v is chosen to be zero on S_1.

The given natural boundary conditions which apply on the parts S_2 and S_3 are now incorporated so that

$$-\int_S kv \frac{\partial u}{\partial n} \, ds = \int_{S_2} vq^* \, ds + \int_{S_3} vh(u - u_\infty) \, ds$$

Thus, with a little rearrangement, the weak variational form becomes: to find u such that $u = u^*$ on S_1 and

$$\int\int_R k\nabla u \cdot \nabla v \, dA + \int_{S_2} vq^* \, ds + \int_{S_3} vh(u - u_\infty) \, ds = \int\int_R fv \, dA$$

or

$$\int\int_R k\nabla u \cdot \nabla v \, dA + h\int_{S_3} uv \, ds = \int\int_R fv \, dA - \int_{S_2} vq^* \, ds + hu_\infty \int_{S_3} v \, ds \quad (6.15)$$

for all v such that $v = 0$ on S_1.

Exercise 6.7

Obtain the weak variational form of the boundary value problem

$$u''(x) + au(x) = -f(x) \qquad 0 < x < 1$$

where $u(0) = 0$, and $u'(1) = bu(1) + c$.

6.7 Language of functionals

In order that the method described above may be generalised to other differential equations, some new terminology must be introduced. In chapter 2 the idea of a functional was developed, in particular for

$$V(u(x)) = \int_0^l \left[\frac{AE}{2} u'^2 + w(u+x) \right] dx \qquad (6.16)$$

where the value of V depends on the choice of the function u. A **linear** functional L satisfies, by definition,

$$L(u+v) = L(u) + L(v)$$

and

$$L(\alpha u) = \alpha L(u)$$

where u, v are any two functions and α is a constant.

Exercise 6.8
Show that

$$L(u(x)) = \int_0^1 [u''(x) + 3u(x)] \, dx$$

is linear.

Exercise 6.9
Show that the functional of (6.16) is not linear.

Many of the integrals of sections 6.5 and 6.6 involve two functions, the trial and the test functions, for example,

$$\int \int_R k\nabla u \cdot \nabla v \, dA \qquad \text{or} \qquad \int_0^1 (u'v' + auv) \, dx$$

A functional depending on two functions may be denoted by $B(u, v)$. It is said to be **bilinear** if it is linear in both u and v. Thus

$$B(u_1 + u_2, v) = B(u_1, v) + B(u_2, v)$$
$$\text{and} \qquad B(\alpha u, v) = \alpha B(u, v), \qquad (6.17)$$

together with similar results for v.

Exercise 6.10
Write down the corresponding results to (6.17) for the second function v.

Exercise 6.11
Show that (a) $\int \int_R k\nabla u \cdot \nabla v \, dA$, and (b) $\int_0^1 (u'v' + auv) \, dx$ are both bilinear, but that $\int_0^1 u^2 v \, dx$ is not.

A functional involving two functions u, v is said to be **symmetric** if u and v can be interchanged without altering the value of the functional, i.e. B is symmetric if

$$B(u, v) = B(v, u).$$

One more definition: $B(u, v)$ is said to be **positive definite** if

$$B(u, u) \geq 0 \quad \text{for all functions } u$$

and

$$B(u, u) = 0 \quad \Longleftrightarrow \quad u \equiv 0$$

Exercise 6.12

Consider the three functionals

$$B_1(u, v) = \int_0^1 (u'v' + uv)\, dx$$

$$B_2(u, v) = \int_0^1 (u'v' - uv)\, dx$$

$$B_3(u, v) = \int_0^1 (u'^3 v' + u^2 v^2)\, dx$$

Show that B_1 and B_2 are symmetric, but B_3 is not; and that B_1, B_3 are positive definite, but B_2 is not.

6.8 Connection between the functionals B, L and V

The weak variational forms, given in equations (6.13), (6.15) and the answer to exercise 6.7, are examples of the general form of the boundary value problem: that u satisfies $u = u^*$ on (part of) the boundary, S_1, and

$$B(u, v) = L(v) \tag{6.18}$$

for all v, $v = 0$ on S_1. Where B and L are functionals and u, v are the trial and test functions.

Once a boundary value problem has been posed in this form, the finite element method may be used directly to approximate to its solution; this will be the subject of the next chapter. In many cases, however, the solution to (6.18) is the same function that minimises a related functional, and so we can establish the link with the finite element method developed in the earlier chapters. In this section the connection is derived and it is shown when the corresponding functional to be minimised exists and how to obtain it. The result will be proved as a theorem. The two approaches, that of satisfying a weak variational form and that of minimising a functional, are put as statements and the proof shows that they are equivalent. (The necessary continuity requirements of the functions for the proof of the theorem will be assumed to be satisfied.) In what follows the functional B is assumed to be bilinear, symmetric and positive definite, and L to be linear.

Statement A: A function u satisfies $u = u^*$ on S_1 and

$$B(u, v) = L(v) \qquad \text{for all } v, v = 0 \text{ on } S_1$$

Statement B: A function u satisfies $u = u^*$ on S_1 and minimises the functional

$$V(u) = \frac{1}{2} B(u, u) - L(u)$$

Theorem: Statements A and B are equivalent.

A preliminary manipulation is needed for the proof.

Consider a particular function $\tilde{u}$ and add to it a (small) multiple ϵ of a second function v, thus forming a new function u which is a perturbation (small change) of $\tilde{u}$,

$$u = \tilde{u} + \epsilon v.$$

If, like the test function, $v = 0$ on the part of the boundary where the essential boundary condition is imposed, both u and $\tilde{u}$ have the same value $(= u^*)$ there. Now consider V evaluated for the function u, using the properties of linearity and symmetry,

$$
\begin{aligned}
V(u) &= V(\tilde{u} + \epsilon v) \\
&= \frac{1}{2} B(\tilde{u} + \epsilon v, \tilde{u} + \epsilon v) - L(\tilde{u} + \epsilon v) \\
&= \frac{1}{2} [B(\tilde{u}, \tilde{u}) + B(\epsilon v, \tilde{u}) + B(\tilde{u}, \epsilon v) + B(\epsilon v, \epsilon v)] - L(\tilde{u}) - L(\epsilon v) \\
&= \frac{1}{2} [B(\tilde{u}, \tilde{u}) + 2\epsilon B(v, \tilde{u}) + \epsilon^2 B(v, v)] - L(\tilde{u}) - \epsilon L(v) \\
&= \frac{1}{2} B(\tilde{u}, \tilde{u}) - L(\tilde{u}) + \epsilon [B(\tilde{u}, v) - L(v)] + \frac{\epsilon^2}{2} B(v, v) \\
&= V(\tilde{u}) + \epsilon [B(\tilde{u}, v) - L(v)] + \frac{\epsilon^2}{2} B(v, v) \qquad (6.19)
\end{aligned}
$$

Now follows the proof.

Proof

To show that statement A implies statement B.

Suppose that $\tilde{u}$ satisfies A, i.e.

$$B(\tilde{u}, v) = L(v) \qquad \text{for all } v$$

Then from (6.19)

$$V(u) = V(\tilde{u}) + \frac{\epsilon^2}{2} B(v, v)$$

Since B is positive definite the term $\frac{\epsilon^2}{2}B(v, v)$ cannot be negative, and is zero only if ϵ or v is zero, in which case $u = \tilde{u}$. So all perturbations of u from $\tilde{u}$ give a larger value of V. So $\tilde{u}$ minimises V.

To show that statement B implies statement A, suppose that $\tilde{u}$ satisfies statement B, i.e. it minimises V. As before

$$V(u) = V(\tilde{u}) + \epsilon[B(\tilde{u}, v) - L(v)] + \frac{\epsilon^2}{2}B(v, v)$$

Suppose that a function is chosen for v and since $\tilde{u}$ is fixed, both B and L, and hence $V(u)$, are numbers. Thus $V(u)$ depends only on ϵ, so we can write

$$V(u) = F(\epsilon).$$

If ϵ is continuously varied, observe that u becomes $\tilde{u}$ when $\epsilon = 0$; thus $V(u) = F(\epsilon)$ assumes its minimum value when $\epsilon = 0$, i.e.

$$\frac{\mathrm{d}F}{\mathrm{d}\epsilon} = 0 \quad \text{when} \ \epsilon = 0.$$

Returning to (6.19) and differentiating with respect to ϵ,

$$\frac{\mathrm{d}F}{\mathrm{d}\epsilon} = \frac{\mathrm{d}V}{\mathrm{d}\epsilon} = [B(\tilde{u}, v) - L(v)] + \epsilon B(v, v),$$

which must be zero when $\epsilon = 0$. Thus

$$B(\tilde{u}, v) - L(v) = 0 \qquad (6.20)$$

No particular function v was chosen, so (6.20) must hold for all v.
The theorem is proved.   □

We are now in a position to show the derivation of the functionals used in the earlier chapters. Consider the boundary value problem with which finite elements were introduced in chapter 2. The extension $u(x)$ of an elastic string satisfies

$$AEu''(x) = -w$$

with boundary conditions $u(0) = 0$ attached at B, and $AE\frac{\mathrm{d}u}{\mathrm{d}x}(l) = T_C$ (the string is extended by a given force, T_C, at C, figure 6.4).

When converted into variational form this becomes: find u such that $u(0) = 0$ and

$$\int_0^l (AEu'' + w)v \, \mathrm{d}x = 0 \qquad \text{for all } v \text{ where } v(0) = 0.$$

On integrating by parts and rearranging,

$$\int_0^l -AEu'v' \, \mathrm{d}x + [AEu'v]_0^l = -\int_0^l wv \, \mathrm{d}x$$

$$\int_0^l -AEu'v' \, \mathrm{d}x + AEu'(l)v(l) - AEu'(0)v(0) = -\int_0^l wv \, \mathrm{d}x$$

B'

C'

T_C

Figure 6.4   An elastic string

When the boundary conditions have been inserted, we have

$$-\int_0^l AEu'v'\,dx + T_C v(l) = -\int_0^l wv\,dx$$

or

$$\int_0^l AEu'v'\,dx - T_C v(l) = \int_0^l wv\,dx$$

Comparing this with the general form $B(u, v) = L(v)$, it is clear that

$$B(u, v) = \int_0^l AEu'v'\,dx \quad \text{and} \quad L(v) = \int_0^l wv\,dx + T_C v(l)$$

Exercise 6.13
Show that B is linear, symmetric and positive definite, and that L is linear.

Since B and L satisfy the conditions of the theorem, the corresponding functional which is minimised is

$$V(u) = \frac{1}{2}B(u, u) - L(u)$$

$$= \frac{1}{2}\int_0^l AEu'^2\,dx - \int_0^l wu\,dx - T_C v(l)$$

$$= \int_0^l \left[\frac{1}{2}AEu'^2 - wu\right]dx - T_C v(l) \tag{6.21}$$

If this is compared with the functional used in chapter 2 (see equation 2.6), the difference is in the absence of the term $-\int_0^l wx\,dx$. (The term $T_B u(0)$ is missing because an essential boundary condition, $u(0) = 0$, has been assumed.) The term $-\int_0^l wx\,dx = -w\frac{l^2}{2}$ is a constant. So the two forms, (2.6) and (6.21), differ only by a constant and the minimum of one will occur for the same function which minimises the other.

To conclude the chapter the functional to be minimised corresponding to Laplace's equation is considered. It had been quoted previously, but now the necessary mathematics is available. As the derivation is straightforward it is left as an exercise.

Exercise 6.14

Show that the two functionals B and L corresponding to solving Laplace's equation with various boundary conditions are (see equation 6.15)

$$B(u, v) = \int\int_R k\nabla u \cdot \nabla v\,dA + h\int_{S_3} uv\,ds$$

$$L(v) = \int\int_R fv\,dA - \int_{S_2} vq^*\,ds + hu_\infty \int_{S_3} v\,ds$$

(a) Assuming that both k and h are positive (as they will be in practical problems), show that B is bilinear, symmetric and positive definite and that L is linear.

(b) Obtain the functional V.

(c) Compare with the functional of chapter 5 and reconcile any difference.

General exercises for chapter 6

1. Consider the boundary value problem

    $$u'' + u = -1 \qquad 0 < u < 0 \qquad \text{where } u(0) = u(1) = 0.$$

    (a) Obtain the analytic solution.

    (b) Form the residual $r(x)$ for a one-parameter trial function $g(x) = \alpha x(1 - x)$.

    (c) Choose α so that $r(0.5) = 0$, i.e. so that the residual at the mid-point is zero. This is an example of the **collocation** method by which the residual is made to be zero at a number of points – as many as there are parameters.

    (d) Choose α to minimise

    $$\int_0^1 r^2(x)\,dx.$$

    This is an example of the **least squares** approach.

    (e) Graph both the residual and error for (c) and (d).

2. Consider the differential equation

$$u'' - u = -x \qquad 0 < x < 1$$

with boundary conditions $u(0) = 0, \; u'(1) = 2$.

(a) Form the analytic solution.

(b) Change the problem into weak variational form.

(c) Form two approximate solutions by choosing a linear and quadratic trial function.

(d) Compare the errors in (c).

3. Recast the boundary value problem

$$u''(x) + au'(x) + bu(x) = -f(x) \quad 0 < x < 1$$

$$\text{where} \quad u(0) = u(1) = 0$$

into the weak variational form: find

$$u : u(0) = u(1) = 0 \quad \text{satisfying} \quad B(u, v) = L(v)$$
$$\text{for all test functions} \quad v : v(0) = v(1) = 0.$$

Identify the functionals B and L.

What are the conditions on a and b so that

(a) B is bilinear?

(b) B is symmetric?

(c) B is positive definite?

4. The defection, w, of a beam simply supported at its ends (figure 6.5), satisfies the differential equation

$$\frac{d^2}{dx^2}\left[EI(x)\frac{d^2 w(x)}{dx^2}\right] = f(x)$$

where E is the modulus of elasticity and $I(x)$ is the moment of inertia of a cross-section, which may vary along the length. The applied downward force

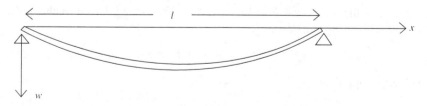

Figure 6.5   A simply supported beam

is $f(x)$. The boundary conditions are that there is no deflection at the ends and the moments are zero there too, i.e.

$$w(0) = w(l) = 0 \quad \text{and} \quad \frac{d^2w}{dx^2}(0) = \frac{d^2w}{dx^2}(l) = 0$$

Convert the problem to weak variational form, and if possible (justifying the process), obtain the corresponding functional to be minimised. Does it make any difference to the variational form if $I(x)$ varies or is constant? What continuity is required of $w(x)$ when the problem is posed in weak variational form? Would piecewise linear trial functions be suitable?

5. (For discussion) Is it possible to build natural boundary conditions into the trial function in a similar way to the essential conditions? If it is possible what would be the advantages? Why is it not done in practice?

6. An insulating wall is constructed of three homogeneous layers of differing conductivity, k_1, k_2, k_3, and widths w_1, w_2, w_3, as shown in figure 6.6.

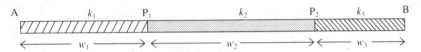

Figure 6.6   Horizontal cross-section of an insulating wall

At the inner side, A, the temperature is given as u^* and at the outer side, B, heat is lost through convection. Expressed as a differential equation, u satisfies

$$k\frac{d^2u}{dx^2} = 0$$

on each layer, with boundary conditions

$$u(A) = u^*, k\frac{du}{dx} \text{ is continuous at the two inner points } P_1, P_2,$$

and $-k_3 du/dx(B) = h(u(B) - u_\infty)$.

Construct the problem in variational form.

7. Obtain the functionals B and L corresponding to the boundary value problem in two-dimensional acoustics:

$$\nabla^2 u + k^2 u = 0 \qquad \text{for } (x, y) \text{ in } R$$

where the boundary conditions are that

$$u = u^* \text{ on } S_1 \quad \text{and} \quad \frac{\partial u}{\partial n} = v^* \text{ on } S_2 \qquad S_1 \cup S_2 = S.$$

Show that B is bilinear and symmetrical, and that L is linear. Can the problem be posed as finding the minimum of a functional?

8. Consider an example in steady state heat flow with the problem variable u and the flow $\mathbf{q} = -k\nabla u$. A region R_1 has prescribed boundary conditions $u = u_1^*$ on part S_1 of the boundary and $q_n = q_1^*$ on the remainder S_1^c. The region R_2 has similar boundary conditions as shown in figure 6.7. The two regions have conductivities k_1, k_2, respectively.

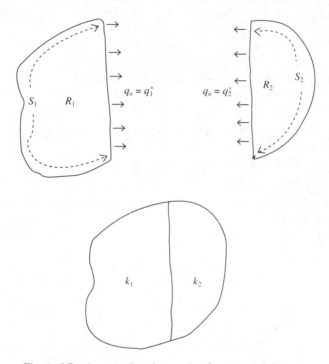

Figure 6.7 A general region made of two materials

Write down the partial differential equation and boundary conditions satisfied by u on R_1 and R_2. Write down also the weak variational form for the two regions.

The two regions are now joined together by making S_1^c and S_2^c common and $q_1^* + q_2^* = 0$. Form the differential equation and boundary conditions, and also the variational form for the combined region. Compare the two forms.

7 The Galerkin approach

7.1 Introduction

The process of transforming a problem from a differential equation to a weak variational form was established in chapter 6. Now, trial functions and, in particular, the finite element form of trial functions are used with the variational form. The end result does not differ from earlier chapters; the same stiffness matrix and equations are derived, but the means of getting there is different. Once the new background ideas have been accepted, the approach may be seen to be elegant and generally applicable. In progressing through the mathematics the discussion will be:

- A little about function spaces and basis functions, particularly in the finite element setting.

- How the Galerkin method leads to the same equations that were obtained previously.

- And hence to see that the more mathematical approach of chapters 6 and 7 has the advantage of starting from the differential equation and may be applied even when a functional to be minimised is not available.

7.2 A little about function spaces

Consider the boundary value problem expressed in the general weak variational form: find a function u satisfying $u = u^*$ on S_1 and

$$B(u, v) = L(v) \qquad \text{for all } v, v = 0 \text{ on } S_1 \tag{7.1}$$

where B is a bilinear, positive definite functional, L is a linear functional and v the test function.

The function u is the trial function. Candidates for the position of 'best' trial function are chosen from a set or class of suitable functions; the class of such functions forms a **function space**. To give an example of a function space, suppose that the shape of u is known to be very smooth, in that the function is continuous and has derivatives of any order. In this case Taylor's theorem states that u in one dimension can be represented by an infinite power series:

$$\sum_{n=0}^{\infty} a_n x^n \tag{7.2}$$

Because of the infinite number of constants, a_n, $n = 0$, 1, $\ldots$, this forms an infinite dimensional function space. In practice, it is easier to limit the size of the function space by considering, say, only polynomials of degree m

$$\sum_{n=0}^{m} a_n x^n \tag{7.3}$$

which form an $m+1$ dimensional function space.

If the continuity requirement placed on u is reduced; say, that it is only piecewise continuous on a finite interval $[0, l]$, then it might be more suitable to consider a (half-range) Fourier series representation

$$\sum_{n=1}^{\infty} b_n \sin n \frac{\pi}{l} x \tag{7.4}$$

Again, this is an infinite dimensional function space, which could be approximated by a truncated series

$$\sum_{n=1}^{m} b_n \sin n \frac{\pi}{l} x \tag{7.5}$$

The finite element method may be thought of as a method for finding the best function when searching through a particular finite dimensional function space. Function spaces may be constructed as a linear combination of **basis** functions. In (7.2) and (7.3) the polynomials $1, x, x^2, \cdots$ are such functions, as are $\sin \frac{\pi}{l} x$, $\sin \frac{2\pi}{l} x, \ldots$ in (7.4) and (7.5).

These ideas are illustrated further in the next section, where we consider the function spaces related to the finite element method and the basis functions which are used to construct them.

7.3 Basis for element function spaces

In chapters 1 and 2 an assumption was made on how u varied on a one-dimensional element $x_i \leq x \leq x_j$ (figure 7.1). We assumed the form $\alpha + \beta x$. This can be constructed from the two basis functions

$$W_1(x) = 1 \qquad\qquad W_2(x) = x$$

or, as was preferred, by the shape functions

$$N_i(x) = \frac{x_j - x}{h} \qquad\qquad N_j(x) = \frac{x - x_i}{h}$$

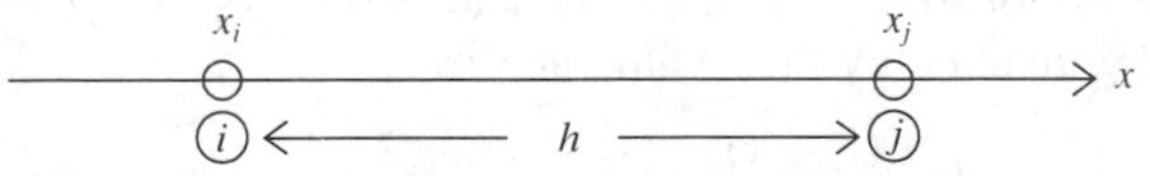

Figure 7.1   Setting for a linear element

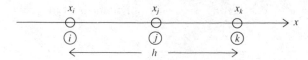

Figure 7.2   Setting for a quadratic element

It is clear that the basis functions are not unique. For example $1 + x$, $1 - x$ also span the same function space.

The reason for the preference for the shape functions comes as a result of the properties

$$N_i(x_i) = 1 \qquad N_i(x_j) = 0$$
$$N_j(x_i) = 0 \qquad N_j(x_j) = 1 \qquad (7.6)$$

If a function u is represented as

$$u(x) = \alpha N_i(x) + \beta N_j(x) \qquad (7.7)$$

then from the properties (7.6) the α and β are the problem variable values at the points x_i and x_j, that is $\alpha = u(x_i)$ and $\beta = u(x_j)$.

In the representation with general basis functions W_1 and W_2,

$$u(x) = \alpha W_1(x) + \beta W_2(x)$$

the values of α and β are unlikely to have any physical significance.

Suppose that, instead of a linear approximation for u, a quadratic is used (figure 7.2):

$$u(x) = \alpha + \beta x + \gamma x^2$$

one set of basis functions are 1, x and x^2. More suited to the finite element method is to develop quadratic shape functions to form a basis. Three nodes are needed, the two end nodes and an additional one in the middle. The properties to be satisfied are an extension of (7.6):

$$N_i(x_i) = 1 \qquad N_i(x_j) = 0 \qquad N_i(x_k) = 0$$
$$N_j(x_i) = 0 \qquad N_j(x_j) = 1 \qquad N_j(x_k) = 0 \qquad (7.8)$$
$$N_k(x_i) = 0 \qquad N_k(x_j) = 0 \qquad N_k(x_k) = 1$$

Consider forming $N_i(x)$ as shown in figure 7.3; since it is zero at x_j and x_k it contains the factors $(x_j - x)$, $(x_k - x)$ and since $N_i(x_i) = 1$, the factors are normalised by dividing by their value at $x = x_i$:

$$N_i(x) = \frac{(x_j - x)}{\left(\frac{h}{2}\right)} \times \frac{(x_k - x)}{(h)} = \frac{2}{h^2}(x_j - x)(x_k - x) \qquad (7.9)$$

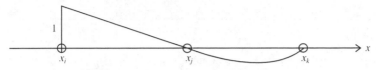

Figure 7.3   The graph of the quadratic N_i

Exercise 7.1
Determine whether the following span the function space of $\alpha + \beta x + \gamma x^2$:

   (a) $x - 1$, $x + 1$, x^2
   (b) 1, $x - 1$, $(x - 1)^2$
   (c) $(x - 1)^2$, $(x + 1)^2$, $x^2 + 1$
   (d) $(x - 1)^2$, $(x + 1)^2$, x^2

Exercise 7.2
Obtain N_j and N_k and sketch their graphs.

Exercise 7.3
Show that if $u(x) = \alpha N_i(x) + \beta N_j(x) + \gamma N_k(x)$ then $\alpha = u(x_i)$, $\beta = u(x_j)$, and $\gamma = u(x_k)$.

Exercise 7.4
Verify that
$$N_i(x) + N_j(x) + N_k(x) \equiv 1.$$

Explain, by using the particular function $u(x) = 1$ in exercise 7.2.

Exercise 7.5
Verify that
$$\frac{\mathrm{d}N_i(x)}{\mathrm{d}x} + \frac{\mathrm{d}N_j(x)}{\mathrm{d}x} + \frac{\mathrm{d}N_k(x)}{\mathrm{d}x} \equiv 0$$

Explain.

7.4 Problem domain function spaces

In section 7.3 we considered making a linear or quadratic approximation on an individual element, but the domain for the problem is the combined domain of the elements. This is reflected in the way that the global approximation to the problem variable $u(x)$ is formed from the local, element approximations. The global approximation is constructed to be continuous; it will be piecewise linear if the approximation is linear on each element or, following the development of section 7.3, it could be piecewise quadratic.

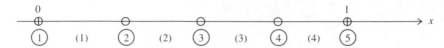

Figure 7.4   The problem domain with four elements

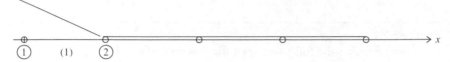

Figure 7.5   The   basis function associated with node 1, ϕ_1

The next two subsections develop sets of basis functions for the global approximation.

7.4.1 Forming a piecewise linear global approximation

We illustrate with a problem domain $[0, 1]$ divided into four linear elements (figure 7.4).

What is a suitable set of linear basis functions to span this function space? The finite element method has been developed using shape functions defined on each element, but not considered outside the element. Global basis functions defined on the whole problem domain are constructed from the element shape functions and inherit their convenient properties. They are associated one with each node and they take the value 1 at 'their' node and 0 at all the others.

The function space is a five-dimensional piecewise linear approximation, being defined by five values – say by the values of u at nodes 1, 2, 3, 4 and 5. What is a suitable set of basis functions to span this function space? Consider, $\phi_1(x)$, the basis function associated with node 1.

In the domain of the first element, $[0, \frac{1}{4}]$, the approximation is linear and it is clearly convenient to choose $\phi_1(x)$ to be the element shape function for node 1, $N_1(x)$. For the rest of the problem domain $\phi_1(x)$ has to be defined so that the resulting function is continuous; the simplest way is to make it zero (figure 7.5). For $\phi_2(x)$ the two shape functions $N_2(x)$ belonging to the first and second elements are used to construct a **hat-shaped** function. Outside these two elements it is defined to be zero, thus giving a continuous, piecewise linear function (figure 7.6). Similarly, $\phi_3(x)$ is constructed from $N_3(x)$ belonging to the two elements 2 and 3 (figure 7.7).

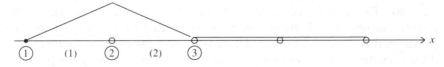

Figure 7.6   The basis function, ϕ_2

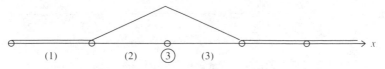

Figure 7.7   The basis function, ϕ_3

Exercise 7.6
Sketch ϕ_4 and ϕ_5.

Note that all the finite element basis functions satisfy

$$\phi_i(\text{node } j) = \begin{cases} 1 & \text{if } i = j \\ 0 & \text{otherwise} \end{cases} \tag{7.10}$$

These properties, which were seen to be part of the definition for shape functions, also hold for the basis functions.

Exercise 7.7
Suppose $u(x) = \alpha_1\phi_1(x) + \alpha_2\phi_2(x) + \alpha_3\phi_3(x) + \alpha_4\phi_4(x) + \alpha_5\phi_5(x)$. Show that $\alpha_i = u(\text{node } i)$.

Clearly ϕ_1, $\phi_2, \dots, \phi_5$ are continuous and piecewise linear and so belong to the function space; also they may be seen to be linearly independent and so span the five-dimensional space. To focus on the meaning of exercise 7.7, consider the expression

$$3\phi_1(x) + 2\phi_2(x) + 4\phi_3(x) - \phi_4(x) + 2\phi_5(x)$$

From the exercise it can be seen that at node 1 its value is 3; and at nodes 2, 3, 4 and 5 it takes the values 2, 4, –1 and 2, respectively. Thus its graph is as shown in figure 7.8.

Exercise 7.8
Sketch the graph of $f(x) = 7\phi_1(x) - 2\phi_2(x) + \phi_3(x) + 4\phi_4(x)$.

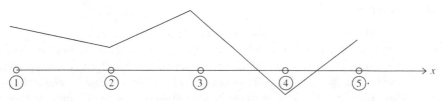

Figure 7.8   The   graph of $3\phi_1(x) + 2\phi_2(x) + 4\phi_3(x) - \phi_4(x) + 2\phi_5(x)$

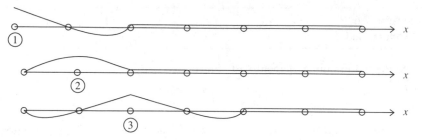

Figure 7.9   The problem domain with three quadratic elements

Figure 7.10 ϕ_1, ϕ_2 and ϕ_3

7.4.2 Forming a piecewise quadratic global approximation

Suppose we use three elements, this time quadratic. Each element will have the three nodes needed to define a quadratic (figure 7.9). Note that it is a seven-dimensional space requiring seven basis functions.

The basis functions are made up from element shape functions shown in figure 7.3 and exercise 7.2, and are continuous piecewise quadratics ϕ_1, ϕ_2 and ϕ_3 (figure 7.10).

Note that ϕ_2, corresponding to a middle node, uses only a shape function for the element, whereas ϕ_3 needs the shape functions from elements 1 and 2 in order to be continuous.

The quadratic basis functions also satisfy the fundamental relations (7.10).

Exercise 7.9

Sketch the remaining four basis functions for piecewise quadratic approximation.

7.4.3 Basis functions for a two-dimensional piecewise linear global approximation

The linear element shape functions for a triangle have been shown previously to be as in figure 7.11. It is left as an exercise to show how these are built up into global basis functions, continuous, piecewise linear and satisfying the property (7.10). Each basis function is again associated with a node and each is shaped like a pyramid. They are non-zero on elements having their node in common and are zero outside these elements.

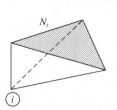
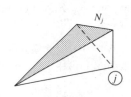
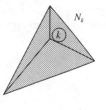

Figure 7.11   Element shape functions

Exercise 7.10
Sketch the basis functions ϕ_i and ϕ_j. See figure 7.12.

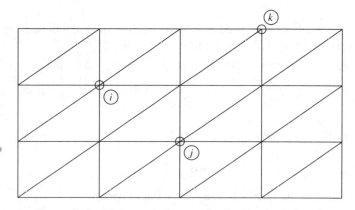

Figure 7.12   Nodes and elements for drawing basis functions

Exercise 7.11
This exercise is concerned with basis functions corresponding to a boundary node k. Sketch $\phi_k(x, y)$.

7.5 Using a general function space

Repeating the variational form, the problem is to find a function u satisfying $u = u^*$ on S_1 and

$$B(u, v) = L(v) \qquad \text{for all } v. \tag{7.11}$$

Suppose that the trial function u is chosen from an N-dimensional function space, spanned by a set of basis functions $W_1, W_2, \ldots, W_N$. Then we set

$$u = \alpha_1 W_1 + \alpha_2 W_2 + \ldots + \alpha_N W_N \tag{7.12}$$

where $\alpha_1, \alpha_2, \ldots, \alpha_N$ are constants yet to be chosen. This is done so that u is a 'good' approximation to the solution. Substituting into (7.11) gives

$$B(\alpha_1 W_1 + \alpha_2 W_2 + \ldots + \alpha_N W_N, v) = L(v)$$

or

$$\alpha_1 B(W_1, v) + \alpha_2 B(W_2, v) + \ldots + \alpha_N B(W_N, v) = L(v) \tag{7.13}$$

since B is linear.

Once a test function v is chosen, the terms $B(W_1, v), B(W_2, v), \ldots,$ and $L(v)$ are constants, so that equation (7.13) becomes a linear algebraic equation for the N unknowns. Thus, to give a solvable set of N equations, N distinct functions, $v_1, v_2, \ldots, n_N$ have to be chosen. Then (7.13) becomes

$$\sum_{c=1}^{N} \alpha_c B(W_c, v_r) = L(v_r) \qquad r = 1, 2, \ldots, N \tag{7.14}$$

The letters r and c have been used to indicate the row and column number in the global matrix structure. This has the form

$$\begin{bmatrix} B(W_1, v_1) & B(W_2, v_1) & \cdots & B(W_N, v_1) \\ B(W_1, v_2) & B(W_2, v_2) & \cdots & B(W_N, v_2) \\ \vdots & \vdots & \ddots & \vdots \\ B(W_1, v_N) & B(W_2, v_N) & \cdots & B(W_N, v_N) \end{bmatrix} \begin{bmatrix} \alpha_1 \\ \alpha_2 \\ \vdots \\ \alpha_N \end{bmatrix} = \begin{bmatrix} L(v_1) \\ L(v_2) \\ \vdots \\ L(v_N) \end{bmatrix}$$

or

$$\mathbf{K\alpha = f} \tag{7.15}$$

where the terms in the matrix $\mathbf{K}$ and vector $\mathbf{f}$ are given by

$$K_{rc} = B(W_c, v_r), \qquad f_r = L(v_r)$$

As a simple illustration consider the boundary value problem

$$u'' + u = -x \qquad 0 < x < 1 \qquad \text{where } u(0) = u(1) = 0 \tag{7.16}$$

Exercise 7.12

Derive the analytic solution

$$u = \frac{\sin x}{\sin 1} - x$$

When expressed in weak variational form, (7.16) becomes the problem of finding u satisfying $u(0) = u(1) = 0$ and

$$\int_0^1 (u'v' - uv) \, dx = \int_0^1 xv \, dx \qquad \text{for all } v \text{ such that } v(0) = v(1) = 0 \tag{7.17}$$

The lowest order polynomial defined on the problem domain and satisfying the two boundary conditions is a quadratic (the finite element approach is not being used here). The function space is one dimensional and a basis is

$$W_1(x) = x(1 - x)$$

setting $u = \alpha_1 W_1$ into (7.17) gives

$$\alpha_1 \left[\int_0^1 (W_1'v' - W_1 v) \, dx \right] = \int_0^1 xv \, dx$$

or

$$\alpha_1 \left[\int_0^1 (1 - 2x)v' - x(1 - x)v \, dx \right] = \int_0^1 xv \, dx$$

The value of α_1 now clearly depends upon the choice of test function v. The conditions $v(0) = v(1) = 0$ restrict the choice of v, and probably the simplest choice is to use the same function that is used for the trial. The calculation is left as an exercise. Another choice for the test function is $\sin \pi x$, which can be seen to satisfy the boundary conditions. In this case α_1 is given by

$$\alpha_1 \left[\int_0^1 (1 - 2x)\pi \cos \pi x - x(1 - x) \sin \pi x \, dx \right] = \int_0^1 x \sin \pi x \, dx$$

which gives, after integration

$$\alpha_1 = \frac{\pi^2}{4(\pi^2 - 1)} = 0.278\ 19...$$

The error in this approximation is

$$e(x) = \frac{\sin x}{\sin 1} - x - 0.278\ 19x(1 - x)$$

or

x	0	0.25	0.50	0.75	1
$e(x)$	0	0.0081	0.0002	0.0079	0

(7.18)

Exercise 7.13

Find α_1 given by (7.17) when v is chosen as $x(1 - x)$. Calculate the values of the error corresponding to (7.18).

Exercise 7.14

Consider the boundary value problem

$$u'' + u = -x \qquad 0 < x < 1 \qquad \text{where } u(0) = 0 \text{ and } u'(1) = 0$$

(a) Obtain the solution

$$u = \frac{\sin x}{\cos 1} - x.$$

(b) Recast the problem in weak variational form. The trial function has now only one boundary value to be built in, that is $u(0) = 0$.

(c) Form an approximate solution in the one-dimensional function space of linear polynomials. Tabulate the error as in (7.18).

(d) Consider the solution in the function space of quadratics and again tabulate the error. [The simplest suitable test functions are x and x^2.]

7.6 The Galerkin method

In all the examples above, the basis functions that form the trial function space could also be used as test functions. The Galerkin method is to do just this, to choose the N basis functions as test functions. Then (7.14) becomes

$$\sum_{c=1}^{N} \alpha_c B(W_c, W_r) = L(W_r) \qquad r = 1, 2, \ldots, N$$

or
$$\mathbf{K\alpha} = \mathbf{f}$$

where the terms in the matrix $\mathbf{K}$ and vector $\mathbf{f}$ are now given by

$$K_{rc} = B(W_c, W_r), \qquad f_r = L(W_r) \tag{7.19}$$

Exercise 7.15
Show that if B is symmetrical, then so is the stiffness matrix $\mathbf{K}$.

7.7 The finite element method

This is the Galerkin method using the finite element basis functions. If $\phi_1, \ldots, \phi_N$ are a set of finite element basis functions, then the trial is made

$$u = \sum_{c=1}^{N} \alpha_c \phi_c. \tag{7.20}$$

Since $\alpha_i = u(\text{at node i}) = u_i$, (7.20) becomes

$$u = \sum_{c=1}^{N} u_c \phi_c.$$

The finite element equations then are

$$\mathbf{Ku} = \mathbf{f}$$

where $K_{rc} = B(\phi_c, \phi_r)$, and $f_r = L(\phi_r)$, $\mathbf{u} = [u_1, u_2, \ldots, u_N]^T$.

Suppose that the problem domain R is divided into E elements and that the functionals B and L are formed by integrals over R. Then from the property of integration

$$
\begin{aligned}
B(u, v) &= \int\int_R F(u, v) \; dA \\
&= \sum_{e=1}^{E} \int\int_{R^e} F(u, v) \; dA \\
&= \sum_{e=1}^{E} B^e(u, v)
\end{aligned}
$$

where F is the particular combination of u, v that forms the functional. Thus

$$
K_{rc} = B(\phi_c, \phi_r) = \sum_{e=1}^{E} K_{rc}^e
$$

where now K_{rc}^e is formed by integrating over the element e.

$$
K_{rc}^e = B^e(\phi_c, \phi_r)
$$

To illustrate this theory and to show how the condensed stiffness matrix fits into these ideas, consider the set of linear basis functions on a one-dimensional domain (figure 7.13).

On any particular element e, all the basis functions are zero except the two, ϕ_i and ϕ_j, related to the two nodes of element. The basis functions are constructed from the element shape functions, in particular for element e, on

$$
x_i \leq x \leq x_j \qquad \phi_i(x) \equiv N_i^e(x) \quad \text{and} \quad \phi_j(x) \equiv N_j^e(x)
$$

Also, the basis functions are non-zero on two neighbouring elements only. So in our example, ϕ_i is only non-zero on elements e and $e - 1$. It can be seen that

$$
B^e(\phi_c, \phi_r) = \begin{cases} B^e(N_c^e, N_r^e) & \text{if each of } r \text{ and } c \text{ is either } i \text{ or } j \\ 0 & \text{otherwise} \end{cases} \tag{7.21}
$$

Thus

$$
\mathbf{K}^e = \begin{array}{c} \\ 1 \\ \\ i \\ j \\ \\ N \end{array}
\begin{array}{c} \begin{array}{ccccccc} 1 & \ldots & i & j & \ldots & N \end{array} \\
\left(\begin{array}{ccccccc}
0 & \ldots & 0 & 0 & \ldots & 0 \\
\vdots & \ddots & \vdots & \vdots & \ddots & \vdots \\
0 & \ldots & K_{ii}^e & K_{ij}^e & \ldots & 0 \\
0 & \ldots & K_{ji}^e & K_{jj}^e & \ldots & 0 \\
\vdots & \ddots & \vdots & \vdots & \ddots & \vdots \\
0 & \ldots & 0 & 0 & \ldots & 0
\end{array}\right) \end{array}
$$

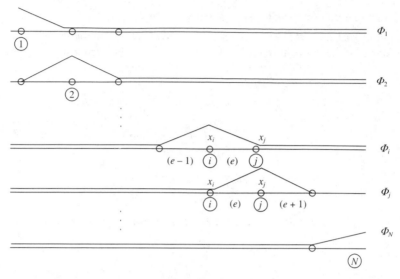

Figure 7.13   The set of basis functions

where $K_{ii}^e = B^e(N_i^e, N_i^e)$ and $K_{ij}^e = B^e(N_j^e, N_i^e)$, etc. The non-zero terms in the matrix form the condensed matrix mentioned in chapters 3 and 4, i.e.

$$\mathbf{K}_c^e = \begin{bmatrix} K_{ii}^e & K_{ij}^e \\ K_{ji}^e & K_{jj}^e \end{bmatrix} \tag{7.22}$$

In a similar manner the finite element vector $\mathbf{f}$ is given by

$$f_r = L(\phi_r) = \sum_{e=1}^{E} L^e(\phi_r) \quad \text{where} \quad L^e(\phi_r) = \begin{cases} L^e(N_r^e) & \text{if } r \text{ is either } i \text{ or } j \\ 0 & \text{otherwise} \end{cases}$$

Similar forms of these equations hold for structures of one, two or three dimensions, and for higher order element approximations.

Exercise 7.16

For Laplace's equation, and a linear element e, with nodes i, j and k, and not experiencing convection, show that

$$K_{rc} = \begin{cases} \iint_{S_e} k \left[\frac{\partial N_c^e}{\partial x} \frac{\partial N_r^e}{\partial x} + \frac{\partial N_c^e}{\partial y} \frac{\partial N_r^e}{\partial y} \right] dxdy & \text{if each of } r \text{ and } c \text{ is either } i, j \text{ or } k \\ 0 & \text{otherwise} \end{cases}$$

Compare with the earlier explanation of chapter 5.

Exercise 7.17

Consider the boundary value problem

$$u''(x) = -1 \qquad 0 < x < 1, \qquad \text{where } u(0) = u(1) = 0$$

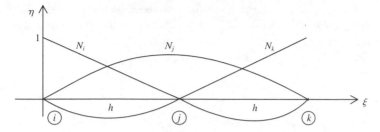

Figure 7.14 Quadratic shape functions with local axes

Form the element stiffness matrix for a

  (a) linear element,

  (b) quadratic element.

[For the quadratic element three shape functions are needed and you may find it convenient to use local axes, as shown in figure 7.14.]

7.8 Broadening the context of Poisson's equation

In chapter 5, Poisson's equation

$$k\left(\frac{\partial^2 u}{\partial x^2} + \frac{\partial^2 u}{\partial y^2}\right) = -f$$

was discussed using the setting of heat flow with various boundary conditions, the most ambitious being convection. The same mathematical equation also models a variety of widely differing physical situations:

- The stress in the bar of arbitrary cross-section in torsion from a twisting torque T (figure 7.15). The problem variable is a stress function ϕ which satisfies

$$\frac{1}{G}\left[\frac{\partial^2 \phi}{\partial x^2} + \frac{\partial^2 \phi}{\partial y^2}\right] = -20 \text{ on } R$$

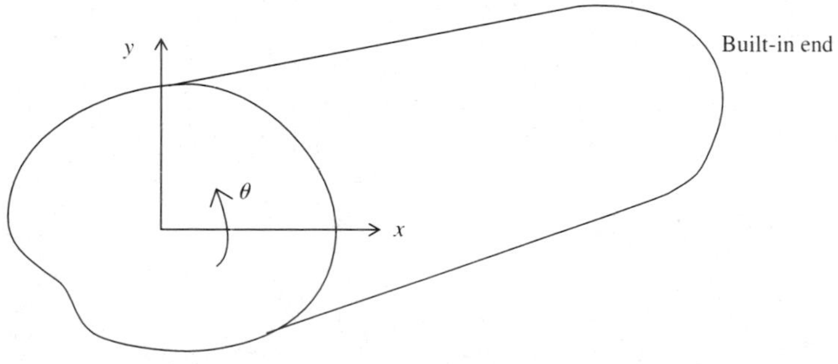

Figure 7.15 Bar of arbitrary cross-section in torsion

and $\phi = 0$ on the boundary of the cross-section; also, if there is a line of symmetry, this corresponds to $\frac{\partial \phi}{\partial n} = 0$. G is the shear modulus of the material and θ is the angle of twist per unit length.

Once ϕ has been found the stresses may be computed from

$$\sigma_{xz} = \frac{\partial \phi}{\partial y}, \qquad \sigma_{yz} = -\frac{\partial \phi}{\partial x}$$

all other stress components are zero. T and θ are related by $T = 2G\theta \int_S \phi \, ds$.

- Ideal, or irrotational fluid flow. This refers to fluid flow which is assumed to be incompressible, and the effect of friction, viscosity, is negligible. These are reasonable assumptions for fluid confined in a channel or around bodies such as weirs, airfoils, buildings, etc., or through earth and dams. There are two possible formulations, using either a velocity potential or a stream function; in both cases the variable satisfies Laplace's equation. The corresponding boundary conditions are shown for the two cases modelling the flow round a circular cylinder between two long flat plates (figure 7.16).

  Velocity potential ϕ, from which the fluid velocity is given by

  $$u = \frac{\partial \phi}{\partial x} \qquad v = \frac{\partial \phi}{\partial y}$$

  Stream function ψ, from which the fluid velocity is given by

  $$u = \frac{\partial \psi}{\partial y} \qquad v = -\frac{\partial \psi}{\partial x}$$

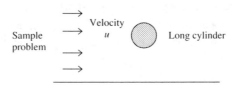

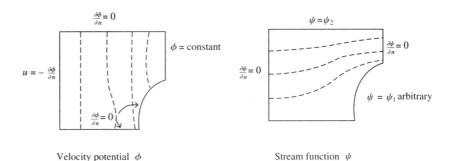

Figure 7.16   The velocity potential and stream function formulation

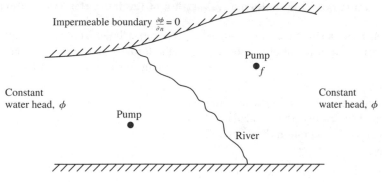

Figure 7.17   An example of groundwater flow

- Groundwater flow or seepage – the pressure in a fluid seeping through porous material. The problem variable is water pressure, the piezometric head (meters), measured from a reference level, and f is the pumping rate $[m^3/(day\ m^3)]$, negative if water is being pumped out. The permeability of the material is $k[m/day]$. Some simple boundary conditions are shown in figure 7.17.

- Electrostatic potential – calculation of capacitance. The electrostatic potential ϕ, in a region empty of charge, satisfies Laplace's equation. Consider, for example, the potential in the region between two non-concentric cylindrical conductors whose cross-section is shown in figure 7.18. Once ϕ has been computed, the capacitance may be obtained from

$$\text{Capacitance} = \epsilon \int \int_S |\nabla\phi|^2 \, ds$$

where ϵ is the dielectric constant using a potential difference of 1 volt.

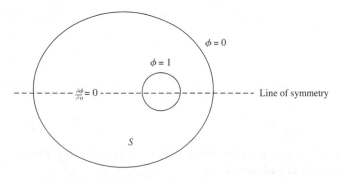

Figure 7.18   An example of electrostatic potential

7.9 Summary of the mathematical setting of the finite element method

Given a boundary value problem in the form of a differential equation,

introduce a trial function u and test function v. Apply the fundamental lemma and integration by parts to obtain

the weak variational form: u satisfies $u = u^*$ on S_1 and $B(u, v) = L(v)$ for all v such that $v = 0$ on S_1.

If u belongs to a finite dimensional function space with basis functions $W_1, W_2, \ldots, W_N$, i.e. $u = \sum_1^N \alpha_c W_c$ and if B is linear in u then the coefficients α_c are given by:

$$\sum_{c=1}^N B(W_c, v)\alpha_c = L(v)$$

for all v, $v = 0$ on S_1.

The Galerkin method chooses $v = W_1, W_2, \ldots, W_N$ giving:

$$\mathbf{K\alpha = f}$$

where, $K_{rc} = B(W_c, W_r)$ and $f_r = L(W_r)$.

Using finite element basis functions $\phi_1, \phi_2, \ldots, \phi_N$ gives,

$$\mathbf{K_G u = f_G}.$$

Given a boundary value problem in the form: u satisfies $u = u^*$ on S_1 and minimises a functional $V(\mathbf{u})$,

the functional V is related to the weak variational form by:

$$V(u) = \tfrac{1}{2}B(u, u) - L(u)$$

if B is bilinear, symmetric and positive definite and L is linear.

Use the finite element basis functions $\phi_1, \phi_2, \ldots, \phi_N$ so that:

$$u = \sum_{c=1}^N u_c \phi_c.$$

Then u satisfies $u = u^*$ on S_1 and:

$$\frac{\partial V}{\partial \mathbf{u}} = 0 \implies \mathbf{K_F u = f_F}.$$

If B is bilinear and positive definite, $\mathbf{K_G = K_F}$, and if L is linear $\mathbf{f_G = f_F}$, i.e. the equations are identical.

General exercises for chapter 7

1. Convert the boundary value problem

$$u'' + u = x \qquad 0 \le x \le 1$$

where $u(0) = 0$ and $u'(1) = u(1) + 1$ into weak variational form. Apply the Galerkin method to obtain an approximate quadratic solution.

2. In weak variational form the deflection, $w(x)$, of a simply supported beam of length l under a loading $f(x)$ satisfies $w(0) = w(l) = 0$ and

$$\int_0^l EIw''(x)v''(x)\,dx = \int_0^l f(x)v(x)\,dx \quad \text{for all} v(x),\ v(0) = v(l) = 0.$$

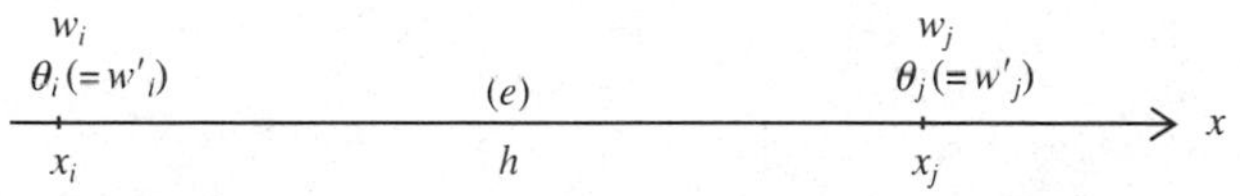

Figure 7.19   An element allowing for continuity of the first derivative

Because of the presence of second derivatives it is necessary for the first derivatives of the trial and test function to be continuous. This can be achieved by introducing a variable $\theta(x) = w'(x)$ at the nodes, i.e. w' (at node i) $= \theta_i$ (figure 7.19). The approximation for w requires a cubic $\alpha x + \beta x^2 + \gamma x^3$ satisfying

$$w(\text{node } i) = w_i \qquad\qquad w'(\text{node } i) = \theta_i$$
$$w(\text{node } j) = w_j \qquad\qquad w'(\text{node } j) = \theta_j.$$

Alternatively, a form of cubic shape functions N_i, M_i, N_j, M_j can be used, so that

$$w(x) = N_i(x)w_i + M_i(x)\theta_i + N_j(x)w_j + M_j(x)\theta_j,$$

where
$$N_i(x_i) = 1 \qquad\qquad N_i'(x_i) = 0$$
$$N_i(x_j) = 0 \qquad\qquad N_i'(x_j) = 0$$

and
$$M_i(x_i) = 0 \qquad\qquad M_i'(x_i) = 1$$
$$M_i(x_j) = 0 \qquad\qquad M_i'(x_j) = 0.$$

There are similar defining conditions for N_j and M_j.

(a) Verify that the shape function formulation for w does satisfy the given conditions.

(b) Sketch the two functions $N_i(x)$, $M_i(x)$ and find their equations.

[Hint: If a polynomial satisfies $f(x_i) = f'(x_i) = 0$ then it contains a factor $(x - x_i)^2$.]

(c) Derive the terms of the condensed element stiffness matrix expressing them in terms of N_i, M_i, N_j and M_j.

3. Show directly that when

$$u = \sum_{c=1}^{N} u_c \phi_c$$

is substituted into the functional,

$$B(u, u) \quad \text{becomes} \quad \mathbf{u}^{\mathrm{T}} \mathbf{K} \mathbf{u}$$

where $\mathbf{u}^{\mathrm{T}} = [u_1, u_2, \ldots, u_N]$ and $K_{rc} = B(\phi_c, \phi_r)$.

Deduce that if the functional B is positive definite, then so is the resulting matrix $\mathbf{K}$.

4. The finite element method using three-noded triangles with arbitrary dimensions when applied to Laplace's equation

$$\nabla^2 \phi = 0$$

produces an element stiffness matrix

$$K_{rc}^e = \begin{cases} \iint_{R^e} \{\nabla N_c^e \cdot \nabla N_r^e\} \, dA & \text{if each of } r, c \text{ is a node} \\ & \text{number of the element} \\ \\ 0 & \text{otherwise} \end{cases}$$

Show that

(a) K^e is symmetric.

(b) The rows (and columns) sum to zero.

Consider whether these results hold for six-noded triangles and for elements of another shape. Also consider in what circumstances the global stiffness matrix is singular. (Assume that the set of M shape functions of a general M-noded element satisfy the property $\sum_{i=1}^{N} N_i = 1$.)

5. Consider, in the same way as in exercise 3, a general linear element with Helmholtz's equation

$$(\nabla^2 + k^2)\phi = 0$$

which produces an element stiffness matrix

$$K_{rc}^e = \begin{cases} \int\int_{R^e}\{\nabla N_c^e \cdot \nabla N_r^e - k^2 N_c^e N_r^e\}\,dA & \text{if each of } r, c \text{ is a node} \\ & \text{number of the element} \\ \\ 0 & \text{otherwise} \end{cases}$$

Show that

(a) K^e is symmetric.

(b) The rows (and columns) sum to $-k^2 A/3$, where A is the area of the triangle.

Consider whether these results hold for six-noded triangles and for elements of another shape.

6. Form the element stiffness matrix for a two-noded linear element corresponding to the boundary value problem

$$u''(x) + au'(x) + bu(x) = f(x) \qquad 0 < x < 1$$

where

$$u(0) = u(1) = 1.$$

Investigate the row and column sums of the element stiffness matrix for the linear element corresponding to the individual terms of the equation above. Consider also whether the sums are different for a quadratic element.

7. An example to calculate the torsion in a twisted bar.

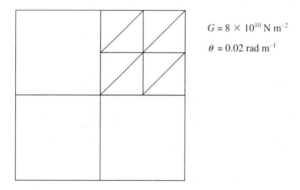

$G = 8 \times 10^{10}$ N m^{-2}

$\theta = 0.02$ rad m^{-1}

Figure 7.20   Cross-section of a twisted bar

The square shaft of side length 0.1 m undergoes a torque which twists it through an angle 0.02 radians per unit length. By analysing a quarter section shown in figure 7.20, obtain ϕ at the nodes and hence estimate σ_{xz}, σ_{yz} for each element.

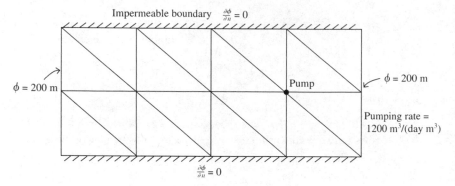

Figure 7.21   Groundwater flow example

8. An example of groundwater flow.
    The rectangular aquifer shown in figure 7.21 has dimensions 2 km × 4 km, is
    bounded on the long sides by an impermeable material and on the short sides
    by a constant piezometric head of 200 m. A river flows through the aquifer
    losing water to the region at the rate of $2\sqrt{2} \times 10^{-1}$ m^3/(day m). Obtain the
    piezometric head at the nodes, and sketch in lines of constant pressure (10,
    say) and indicate the direction of the flow.

8 Element computation

8.1 Introduction

A very significant strength of the finite element method is that it can, to a large extent, be automated. The discretisation of the problem domain, the forming of the element matrices and their combining into global matrices, the solving of the equations and the displaying of the solution, are all tasks which can be adapted for the computer. In this chapter the technique for computing the terms of the element matrices will be described. At the heart of the process is a **mapping**, which relates the actual element to an element of more regular shape and with axes in a more convenient position. Added to this is a numerical technique for the integration, once the terms of the matrix have been expressed as integrals on the regular shape. Finally, a brief description is given on how the discretisation of the problem domain, the mesh generation, may be automated. Thus the chapter describes:

- The idea of a mapping between an actual element and a corresponding idealised element, called the **master element**, and how the mapping is constructed.
- How the computation on the actual element is expressed as an integral on the master element.
- The numerical technique for integrating on the master element.
- The concept of mesh generation and how a mapping provides one approach.

A model problem is chosen to demonstrate how the idea of a mapping arises naturally and to introduce the relevant techniques.

8.2 A simple illustrative example

The terms of the element stiffness matrices and the force vectors are integrals which are not too difficult to evaluate for one-dimensional problems, but they do pose serious problems in two or three dimensions when general elements are being analysed. Imagine trying to integrate on a triangular or quadrilateral element with curved sides. A mapping comes to the rescue by changing the scene of the action to a regular shape for which there are well-established rules for numerical integration. Although the computation for a one-dimensional element does not really present a problem, it can provide the setting for developing a general technique.

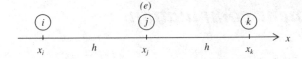

Figure 8.1   A quadratic element

8.2.1 The illustrative problem

Consider

$$-u''(x) + u(x) = x, \tag{8.1}$$

and suppose quadratic elements are to be used. A general element is shown in figure 8.1, with nodes i, j and k equally spaced.

To work on a particular integral, consider the i, jth term in the element stiffness matrix,

$$K_{ij}^e = \int_{x_i}^{x_k} \left[N_j'(x)N_i'(x) + N_j(x)N_i(x) \right] dx, \tag{8.2}$$

where

$$N_i(x) = \frac{1}{2h^2}(x_j - x)(x_k - x) \quad N_i'(x) = \frac{1}{2h^2}(2x - x_j - x_k)$$

and

$$N_j(x) = \frac{1}{h^2}(x - x_i)(x_k - x) \quad N_j'(x) = \frac{1}{h^2}(x_i + x_k - 2x) \tag{8.3}$$

There is no real difficulty here, but the integrand can be slightly simplified by a change of variable

$$\xi = x - x_j$$

i.e. by moving the origin to x_j. Or better still, by adding a scaling factor so that the integration takes place on $[-1, 1]$ rather than on $[x_i, x_k]$,

$$\xi = \frac{1}{h}(x - x_j)$$

Rearranging

$$x = x_j + h\xi. \tag{8.4}$$

A way to view (8.4) is as a mapping from the interval $[-1, 1]$, now thought of as the master element shown in figure 8.2, onto the actual element. This mapping is

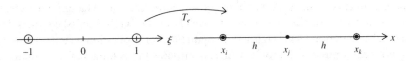

Figure 8.2   Mapping from a quadratic master element

called T_e where e is the element onto which the master element is to be mapped.

$$T_e : \; x = x_j + h\xi. \tag{8.5}$$

We now look in more detail at the process of transferring the various items of the integral (8.2) onto the master element. It will be broken down into small steps, which for this particular integral are rather trivial, but each step corresponds to a separate topic in the structure of a general integration procedure which will be discussed later.

8.2.2 What happens to the shape functions?

Consider for example the particular shape function $N_i(x)$, and apply the mapping (8.5). This is simply a change of variable which results in a function of ξ, say F, given by

$$F(\xi) = N_i(x = x_j + h\xi) = \frac{1}{2h^2}(x_j - x_j - h\xi)(x_k - x_j - h\xi)$$

$$= \frac{1}{2h^2}(-h\xi)(h - h\xi)$$

$$= \frac{1}{2}\xi(\xi - 1)$$

The function $F(\xi)$ may be recognised as the shape function of the master element corresponding to the point $\xi = -1$ and will be called $\hat{N}_i^Q$, using the hat symbol ' $\hat{}$ ' to refer to the master element, and the Q in the superscript to indicate that the shape function is a quadratic. (This latter addition makes the notation increasingly clumsy, but it will be used temporarily to make a distinction between the quadratic and linear shape functions belonging to the same element.) The result, that

$$N_i^Q(x(\xi)) = \hat{N}_i^Q(\xi),$$

is significant, as it implies that the shape function on the actual element is to be replaced by the corresponding master element shape function in the transformed integral. This general approach, of working with the shape functions of the master element, is fundamental to the calculations. The extent to which this holds for a general mapping is discussed in section 8.6.4. To repeat, in the stiffness integral (8.2)

$$N_i^Q(x) = \frac{1}{2h^2}(x_j - x)(x_k - x) \quad \text{becomes} \quad \hat{N}_i^Q(\xi) = \frac{1}{2}\xi(\xi - 1) \tag{8.6}$$

Exercise 8.1
Show that

$$N_j^Q(x) = \frac{1}{h^2}(x - x_i)(x_k - x) \quad \text{becomes} \quad \hat{N}_j^Q(\xi) = 1 - \xi^2.$$

8.2.3 *What happens to differentiation with respect to* x?

The derivatives in (8.2) must be replaced with derivatives with respect to ξ. This is achieved by using the chain rule and equation (8.4) as,

$$\frac{d}{dx} = \frac{d}{d\xi}\frac{d\xi}{dx}$$

$$= \frac{1}{h}\frac{d}{d\xi} \quad \text{using (8.4)}$$

8.2.4 *What happens to* 'dx'?

Similarly, the element of length dx is related to $d\xi$ by

$$dx = h\,d\xi \tag{8.7}$$

Here h is a magnification factor resulting from the mapping. In two dimensions, it is the Jacobian to be described later in sections 8.6.2–3 and 8.6.6.

8.2.5 *The new integral and its evaluation*

Bringing all terms derived in sections 8.2.2–8.2.4 together, we can obtain a new integral entirely in terms of ξ; in other words, the computation of the stiffness matrix has been transferred from the actual element to the master element. Now the i, jth term reads,

$$K_{ij}^e = \int_{-1}^{1}\left[\frac{1}{h}\hat{N}_j'(\xi)\frac{1}{h}\hat{N}_i'(\xi) + \hat{N}_j(\xi)\hat{N}_i(\xi)\right]h\,d\xi, \tag{8.8}$$

and since $\hat{N}_j(\xi) = (1 - \xi^2)$, $\hat{N}_i(\xi) = \frac{\xi}{2}(\xi - 1)$,

$$= \int_{-1}^{1}\left[\frac{1}{h}(-2\xi)(\xi - \frac{1}{2}) + \frac{h}{2}(1 - \xi^2)\xi(\xi - 1)\right]d\xi$$

$$= \int_{-1}^{1}\left[-\frac{1}{h}(2\xi^2 - \xi) - \frac{h}{2}(\xi - \xi^2 - \xi^3 + \xi^4)\right]d\xi \tag{8.9}$$

and noting that

$$\int_{-1}^{1}\xi^n d\xi = \begin{cases} 0 & \text{if } n \text{ is odd} \\ \frac{2}{n+1} & \text{if } n \text{ is even} \end{cases} \tag{8.10}$$

we obtain finally

$$K_{ij}^e = -\frac{4}{3h} + \frac{2h}{15}$$

Exercise 8.2
Transfer the following term from the element force vector onto the master element

$$f_j^e = \int_{x_i}^{x_k} x N_j(x)\,dx$$

To evaluate the integral numerically, one approach is to use a Newton–Cotes formula, say the trapezoidal rule or Simpson's rule, which requires evenly spaced data. However, a more convenient approach is to use Gaussian quadrature which requires function evaluations at carefully chosen points and thereby gains in accuracy. An outline of the method and a list of some relevant formulae are given in the next section, but for this immediate example the Gaussian three-point formula is suitable, being exact for any polynomial of order five or less. The rule approximates the integral of a general function $f(x)$ on the interval -1 to 1 by,

$$\int_{-1}^{1} f(x)\,dx \approx \frac{5}{9}f(-\sqrt{0.6}) + \frac{8}{9}f(0) + \frac{5}{9}f(\sqrt{0.6}) \qquad (8.11)$$

Applying this to (8.9) gives

$$
\begin{aligned}
K_{ij}^e = &\frac{5}{9}\left[-\frac{1}{h}\left[(2(-\sqrt{0.6})^2) - (-\sqrt{0.6})\right]\right] \\
&+ \frac{5}{9}\left[-\frac{h}{2}\left[(-\sqrt{0.6}) - (-\sqrt{0.6})^2 - (-\sqrt{0.6})^3 + (-\sqrt{0.6})^4\right]\right] \\
&+ \frac{8}{9}\left[-\frac{1}{h}[2(0^2) - 0] - \frac{h}{2}[(0 - (0)^2 - (0)^3 + (0)^4)]\right] \\
&+ \frac{5}{9}\left[-\frac{1}{h}\left[2(\sqrt{0.6}^2) - \sqrt{0.6}\right]\right] \\
&+ \frac{5}{9}\left[-\frac{h}{2}\left[(\sqrt{0.6}) - (\sqrt{0.6})^2 - (\sqrt{0.6})^3 + (\sqrt{0.6})^4\right]\right] \\
= &\frac{5}{9}\left[-\frac{1}{h}(4(0.6)) \quad \frac{h}{2}(-2(0.6) + 2(0.6)^2)\right]
\end{aligned}
$$

$$\text{eventually...} = -\frac{4}{3h} + \frac{2h}{15} \quad \text{(as before)}$$

The above computation may look significantly more complicated than the direct integration, but the virtue is that it is easily programmable and the arithmetic can be left to the computer.

Having now illustrated that the integrals which are the terms of the stiffness matrix can be expressed in terms of the master element variable and shape functions, we now ask about the mapping itself. Can this be conveniently set up using the master element shape functions?

8.2.6 The mapping formed by shape functions

The mapping between x and ξ was easily formed in section 8.2.1 by means of a shift followed by a scaling. Generally, however, it is straightforward to use the master element shape functions as a means of interpolation. The mapping function $x = g(\xi)$ is to be chosen so that when $\xi = -1$, then $x = x_i$ and when $\xi = -1$, then $x = x_k$ (and incidentally because it is a linear mapping the midpoint of the master element $\xi = 0$ maps to the mid-point of the actual element $x = x_j$). This process of forming a linear expression which assumes given values at two given points can be carried out, as was seen in section 3.4.1, using shape functions. Quoting the earlier result, the same mapping (8.4) can be expressed as

$$x = \hat{N}_i^L(\xi)x_i + \hat{N}_k^L(\xi)x_k \qquad \text{where } \hat{N}_i^L = \frac{1}{2}(1-\xi), \hat{N}_k^L = \frac{1}{2}(1+\xi) \qquad (8.12)$$

where, because the interpolation is linear, linear shape functions are used. It is interesting to note that a quadratic mapping could have been used for the geometry with the three points $\xi = -1, 0, 1$ mapped to $x = x_i, x_j, x_k$, respectively, but it can be shown to be the same as the linear mapping if x_j is the mid-point of x_i, x_k.

Exercise 8.3
Show that if $x_j = \frac{1}{2}(x_i + x_k)$ then

$$x = \hat{N}_i^Q(\xi)x_i + \hat{N}_j^Q(\xi)x_j + \hat{N}_k^Q(\xi)x_k$$

is the same mapping as (8.12).

Note that shape functions are used for the mapping and shape functions are also used for the problem variable representation, and they need not be the same. If they are the same, then the element is called **isoparametric**, but if the order for the mapping is lower than that of the problem variable then the element is **subparametric**, and otherwise it is said to be **superparametric**. The example given was set up with a linear mapping and quadratic problem variable and so is subparametric (figure 8.3). As shown in the exercise, the mapping could also be thought of as quadratic and from this point of view the element is isoparametric.

8.2.7 A revision of the key ideas

The procedure outlined above has been developed by means of a one-dimensional example, but it holds more generally for elements in two or three dimensions. It results in the computation of the stiffness matrix taking place on a master element, with the link between the actual element and the master element being a mapping. All the terms of the stiffness matrix, although conceived on the actual matrix, can be formed using the **master element shape functions**. In

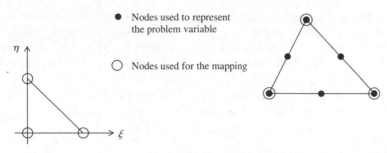

Figure 8.3 A subparametric element

particular, the shape functions of the master element are used in forming:

- the mapping itself,
- the shape functions on the actual element,
- the derivatives,
- the change in area (used in the integral).

Finally, the integral itself, now expressed on the master element, is evaluated numerically, usually by Gaussian quadrature. The numerical integration rules will be given for reference.

8.3 Gaussian quadrature

The basic idea behind Gaussian quadrature, compared with the Newton–Cotes approach, is to trade the even spacing of the positions of the evaluation points for an increase in accuracy.

8.3.1 On the interval [–1,1]

The general formula using n points for one dimension is

$$\int_{-1}^{1} f(x)\, dx \approx \sum_{i=1}^{n} w_i f(x_i) \tag{8.13}$$

where x_i are n evaluation points, known as Gauss points, and w_i are the corresponding weights. The interval for integration is standardised to be $[-1, 1]$, but this is not a restriction because a general interval can be mapped to it using the procedure described above.

For higher dimensions a similar approach is used

$$\int\int_{R} f(x, y)\, dA \approx \sum_{i=1}^{n} w_i f(P_i)$$

where P_i are selected points in the region of integration R.

The points are chosen specifically to gain accuracy rather than to be evenly spaced. An example of how this can be achieved is shown in general exercise 1 at the end of the chapter. The criterion used to determine the position of the points and the weights is to make the formula exact for as high an order polynomial as possible. In one dimension an n-point formula can be made to be exact for a polynomial of order $2n - 1$; thus, in our example the three-point formula was needed to obtain an exact result for the quartic (or for a quintic).

Exercise 8.4

Evaluate directly and by using the appropriate Gauss formula

$$\int_{-1}^{1} (x^5 - 3x^2 + 4)\, dx$$

8.3.2 *On the standard square*

The standard square is taken to be $[-1, 1] \times [-1, 1]$. The quadrature formulae are a direct development of those given in figure 8.4. They are obtained by repeated use of (8.13) for one variable at a time, as

$$
\begin{aligned}
\int_{-1}^{1}\int_{-1}^{1} f(x, y)\, dx\, dy &= \int_{-1}^{1}\left[\int_{-1}^{1} f(x, y)\, dx\right] dy \\
&\approx \int_{-1}^{1}\left[\sum_{i=1}^{n} w_i f(x_i, y)\right] dy \\
&\approx \sum_{j=1}^{n} w_j \left[\sum_{i=1}^{n} w_i f(x_i, y_j)\right] \\
&= \sum_{i=1}^{n}\sum_{j=1}^{n} w_i w_j f(x_i, y_j).
\end{aligned}
\tag{8.14}
$$

n	Points	Weights
1	0.0	2
2	$\pm\frac{1}{\sqrt{3}}$	1
3	$\pm\sqrt{0.6}$	$\frac{5}{9}$
	0	$\frac{8}{9}$

Figure 8.4   Gaussian quadrature on $[-1, 1]$

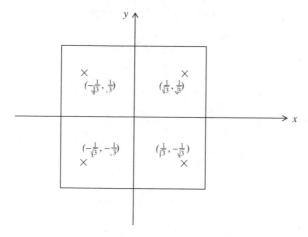

Figure 8.5 Positions of the Gauss points for the four-point rule on the standard square

Thus, for example, the four-point rule (figure 8.5) is

$$\int_{-1}^{1}\int_{-1}^{1} f(x, y)\, dx\, dy \approx f(\frac{1}{\sqrt{3}}, \frac{1}{\sqrt{3}}) + f(-\frac{1}{\sqrt{3}}, \frac{1}{\sqrt{3}})$$
$$+ f(-\frac{1}{\sqrt{3}}, -\frac{1}{\sqrt{3}}) + f(\frac{1}{\sqrt{3}}, -\frac{1}{\sqrt{3}})$$

Exercise 8.5
Evaluate directly and by using the appropriate Gauss formula

$$\int_{-1}^{1}\int_{-1}^{1} (x^3 + 3x^2 y^2)\, dx\, dy$$

8.3.3 On the standard triangle

The standard triangle is taken to be right-angled, isosceles with the equal sides having unit length as shown in figure 8.6. The most commonly used quadrature rules in this region are shown in figure 8.7.

Exercise 8.6
Evaluate directly and by using the appropriate Gauss formula

$$\iint_{R} (x^2 + 3xy)\, dx\, dy$$

where R is the isosceles right-angled triangle formed by $(0, 0), (1, 0)$ and $(0, 1)$.

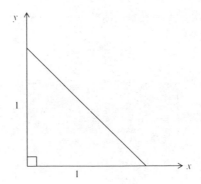

Figure 8.6   The standard triangle

Order	Points	Weights
Linear	$(\frac{1}{3}, \frac{1}{3})$	$\frac{1}{2}$
Quadratic	$(\frac{1}{2}, 0)$	$\frac{1}{6}$
	$(\frac{1}{2}, \frac{1}{2})$	$\frac{1}{6}$
	$(0, \frac{1}{2})$	$\frac{1}{6}$
Cubic	$(0, 0)$	$\frac{3}{120}$
	$(1, 0)$	$\frac{3}{120}$
	$(0, 1)$	$\frac{3}{120}$
	$(\frac{1}{2}, 0)$	$\frac{8}{120}$
	$(\frac{1}{2}, \frac{1}{2})$	$\frac{8}{120}$
	$(0, \frac{1}{2})$	$\frac{8}{120}$
	$(\frac{1}{3}, \frac{1}{3})$	$\frac{27}{120}$

Figure 8.7   Gaussian quadrature on the triangle [(0,0), (1,0), (0,1)]

8.4 Master elements and corresponding shape functions

Shape functions form the basis, in both the technical and the colloquial sense, for constructing a mapping and for representing a problem variable.

Essentially, the shape functions provide a convenient means of producing a range of polynomial approximations which return given function values at given points and interpolate smoothly in between. In the search for greater accuracy, elements with more nodes and different shapes are introduced, which implies using higher order polynomials in the interpolation.

The simplest interpolation in two dimensions is given by a linear polynomial, i.e. one containing three constants a_1, a_2, and a_3,

$$g(\xi, \eta) = a_1 + a_2\xi + a_3\eta \tag{8.15}$$

The shape functions are a set of linearly independent functions which span (can be used to produce any member of) a polynomial function space. Each shape function is associated with a node and the definition of each function is that it should belong to the space, and take the value 1 at its own node and zero at each of the others, i.e.

$$\hat{N}_i(\hat{P}_j) = \begin{cases} 1 & \text{if } j = i \\ 0 & \text{if } j \neq i \end{cases} \tag{8.16}$$

This property ensures linear independence and has the significant bonus, mentioned before, that the interpolation for a function g at a general point $\hat{P}$ is

$$g(\hat{P}) = \sum_i g_i \hat{N}_i(\hat{P})$$

where $g_i = g(\hat{P}_i)$.

In general, there must be as many shape functions and nodes as arbitrary constants a_1, a_2, a_3, ... in the interpolating polynomial. For instance, in the linear case given in equation (8.15), there are three nodes and three corresponding shape functions.

Note that the three constants of (8.15) could have been given through using a different expression, such as

$$G(\xi, \eta) = a_1 + a_2\xi + a_3\xi\eta$$

but this obviously looks strange, missing out the linear term in η and is, in this sense, incomplete. The sensible approach is to take terms from the top of the pyramid in figure 8.8, adding successive rows to give polynomials of increasing degree. This will be illustrated with the different elements considered later.

$$1$$

$$\xi \quad \eta$$

$$\xi^2 \quad \xi\eta \quad \eta^2$$

$$\xi^3 \quad \xi^2\eta \quad \xi\eta^2 \quad \eta^3$$

Figure 8.8 Pyramid of polynomial terms

If a term or terms are missed out, the polynomial is said to be **incomplete**. A significant example is when the constant term is missing; then, the interpolation will be unable to represent a constant value over the element, i.e. a constant temperature or a constant displacement, which is obviously undesirable.

Some two-dimensional master elements and their shape functions are listed below.

8.4.1 Three-noded master triangle

This triangle has been met before, with the lengths of the sides as l, rather than the unit length used now (figure 8.9). Three nodes indicate that the interpolation is linear and hence the shape functions will also be linear (geometrically three points fix a plane whose equation is linear).

In order to find $\hat{N}_1$ we need a linear expression satisfying the defining property (8.16). This expression can be developed from the equation of the line $\hat{P}_2\hat{P}_3$ i.e. $\xi + \eta - 1 = 0$. The expression $\xi + \eta - 1$ is zero at $\hat{P}_2$ and $\hat{P}_3$, so partly satisfies the required properties. At $\hat{P}_1$, $\xi + \eta - 1 = 0 + 0 - 1 = -1$ and not $+1$ as required : so we divide by -1. Thus $\hat{N}_1(\xi, \eta) = (\xi + \eta - 1)/(-1) = 1 - \xi - \eta$.

Exercise 8.7
Using the equations of $\hat{P}_1\hat{P}_3$ and $\hat{P}_1\hat{P}_2$, show that $\hat{N}_2(\xi, \eta) = \xi$ and $\hat{N}_3(\xi, \eta) = \eta$.

Exercise 8.8
Verify that
$$\hat{N}_1(\xi, \eta) + \hat{N}_2(\xi, \eta) + \hat{N}_3(\xi, \eta) = 1$$

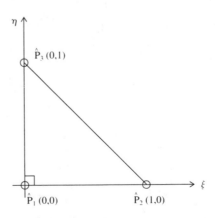

Figure 8.9   The three-noded master triangle

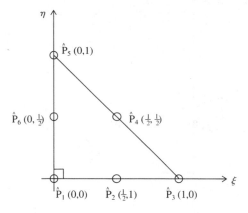

Figure 8.10   Six-noded master triangle

8.4.2 Six-noded master triangle

The six nodes (figure 8.10) indicate quadratic interpolation and hence quadratic shape functions. This is because a quadratic expression of the form

$$a_1 + a_2\xi + a_3\eta + a_4\xi^2 + a_5\xi\eta + a_6\eta^2$$

has six constants, a_1, a_2, ...,a_6, which can be chosen to satisfy six properties (conditions) at the nodes.

To obtain $\hat{N}_1(\xi, \eta)$ we look for a quadratic which again has the defining properties (8.16). Instead of starting with the quadratic expression and determining $a_1, a_2, \ldots, a_6$, we proceed as before by the indirect method of writing down the equations of lines joining suitable nodes (figure 8.11).

$\xi + \eta - 1$ is zero at nodes $\hat{P}_3$, $\hat{P}_4$ and $\hat{P}_5$. At the remaining two nodes, $\hat{P}_2$ and $\hat{P}_6$, $\xi + \eta - \frac{1}{2}$ is zero. Thus the product of these two factors will be zero at all five

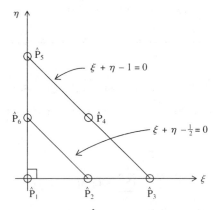

Figure 8.11   Forming $\hat{N}_1$ from line equations

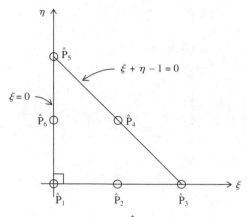

Figure 8.12   Forming $\hat{N}_2$ from line equations

nodes. Hence the polynomial $F_1(\xi, \eta) = (\xi + \eta - 1)(\xi + \eta - \frac{1}{2})$ has only to be modified to give the value 1 at $\xi = \eta = 0$ to become $\hat{N}_1$. Now $F_1(0, 0) = \frac{1}{2}$; so

$$\hat{N}_1(\xi, \eta) = \frac{(\xi + \eta - 1)(\xi + \eta - \frac{1}{2})}{1/2}$$

$$= 2(\xi + \eta - 1)(\xi + \eta - \frac{1}{2})$$

Note that when expanded, $\hat{N}_1(\xi, \eta) = 2\xi^2 + 4\xi\eta + 2\eta^2 - 3\xi - 3\eta + 1$, which is of the quadratic form in ξ and η.

Using the same method (see figure 8.12)

$$F_2(\xi, \eta) = \xi(\xi + \eta - 1), \qquad F_2(\tfrac{1}{2}, 0) = -\tfrac{1}{4}$$

So

$$\hat{N}_2(\xi, \eta) = 4\xi(1 - \xi - \eta)$$

Exercise 8.9

Show that

$$\hat{N}_3(\xi, \eta) = \xi(2\xi - 1) \quad \hat{N}_4(\xi, \eta) = 4\xi\eta$$

$$\hat{N}_5(\xi, \eta) = \eta(2\eta - 1) \quad \hat{N}_6(\xi, \eta) = 4\eta(1 - \xi - \eta)$$

Exercise 8.10

Verify that

$$\sum_{i=1}^{6} \hat{N}_i(\xi, \eta) = 1$$

8.4.3 Four-noded square

The derivation of the shape functions follows the same approach as for the previous elements and is left as an exercise.

Exercise 8.11

Write down the equations of the sides of the square of figure 8.13 and hence show that

$$\hat{N}_1(\xi, \eta) = \frac{1}{4}(1 - \xi)(1 - \eta)$$

$$\hat{N}_2(\xi, \eta) = \frac{1}{4}(1 + \xi)(1 - \eta)$$

$$\hat{N}_3(\xi, \eta) = \frac{1}{4}(1 + \xi)(1 + \eta)$$

$$\hat{N}_4(\xi, \eta) = \frac{1}{4}(1 - \xi)(1 + \eta)$$

(8.17)

Exercise 8.12

Verify that

$$\sum_{i=1}^{4} \hat{N}_i(\xi, \eta) = 1$$

Note that when the expression for each shape function is expanded it has the form

$$\hat{N}_i(\xi, \eta) = a_1 + a_2\xi + a_3\eta + a_4\xi\eta$$

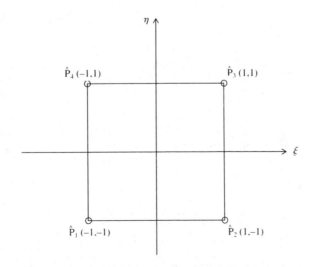

Figure 8.13   The four-noded square master element

and does not include the full range of terms needed for a general quadratic, because terms in ξ^2 and η^2 are missing. It follows that an approximation using these four shape functions is incomplete and will not be as flexible or powerful as that resulting from the six-noded triangle of section 8.4.2.

8.4.4 Eight-noded square master element

It is not too obvious how to proceed, but the method for the four-noded square is used as a start. The product

$$E_1(\xi, \eta) = (1 - \xi)(1 - \eta)$$

gives zero at nodes $\hat{P}_3$, $\hat{P}_4$, $\hat{P}_5$, $\hat{P}_6$ and $\hat{P}_7$ leaving only $\hat{P}_2$ and $\hat{P}_8$ missing. The line through $\hat{P}_2$ and $\hat{P}_8$ is $\xi + \eta + 1 = 0$ (figure 8.14). So, consider the product

$$F_1(\xi, \eta) = (1 - \xi)(1 - \eta)(1 + \xi + \eta)$$

When expanded, $F_1(\xi, \eta) = 1 - \xi^2 - \xi\eta - \eta^2 + \xi^2\eta + \xi\eta^2$, which is a quadratic in each of ξ and η. It may be noticed that there are no linear terms in ξ and η, but other shape functions will provide them.

When suitably scaled, F_1 gives $\hat{N}_1$ as

$$\hat{N}_1(\xi, \eta) = -\frac{1}{4}(1 - \xi)(1 - \eta)(1 + \xi + \eta) \qquad (8.18)$$

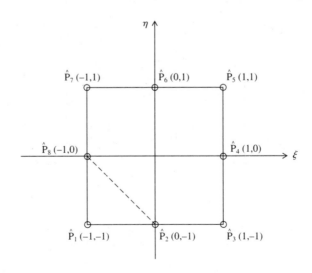

Figure 8.14   The eight-noded square

Exercise 8.13

Show that

$$\hat{N}_2(\xi, \eta) = \frac{1}{2}(1 - \xi^2)(1 - \eta) \tag{8.19}$$

For reference, the other shape functions are

$$\hat{N}_3(\xi, \eta) = \frac{1}{4}(1 + \xi)(1 - \eta)(\xi - \eta - 1) \qquad \hat{N}_4(\xi, \eta) = \frac{1}{2}(1 - \eta^2)(1 + \xi)$$

$$\hat{N}_5(\xi, \eta) = \frac{1}{4}(1 + \xi)(1 + \eta)(\xi + \eta - 1) \qquad \hat{N}_6(\xi, \eta) = \frac{1}{2}(1 - \xi^2)(1 + \eta)$$

$$\hat{N}_7(\xi, \eta) = -\frac{1}{4}(1 - \xi)(1 + \eta)(\xi - \eta + 1) \qquad \hat{N}_8(\xi, \eta) = \frac{1}{2}(1 - \eta^2)(1 - \xi)$$

$$\tag{8.20}$$

8.5 Some element mappings

We now consider the technique for forming a mapping which relates the master element to a corresponding element in the problem domain. This is constructed in such a way that the nodes of the master element are mapped to the corresponding nodes of the actual element. Once set up this way, nodes to nodes, the mapping will more generally map points of the master element $\hat{R}^e$ in the (ξ, η) plane to points of the actual element R^e in the (x, y) plane.

The mapping is a pair of interpolation formulae, i.e. one interpolation that gives the x-coordinates of the nodes and the other which does the same for the ys. The interpolation uses the shape functions which were developed in the previous section. Three examples will suggest a general approach.

8.5.1 Isosceles right-angled triangle mapped onto a general three-noded triangle

Suppose that a mapping T_e is required from the master element onto a general triangle $P_1 P_2 P_3$ in the x,y plane, so that $\hat{P}_1 \rightarrow P_1$, $\hat{P}_2 \rightarrow P_2$, $\hat{P}_3 \rightarrow P_3$ (figure 8.15). As a consequence the sides of $\hat{P}_1 \hat{P}_2 \hat{P}_3$ map onto the sides of $P_1 P_2 P_3$.

The interpolation mapping is given by

$$T_e: \quad \begin{aligned} x &= \hat{N}_1(\xi, \eta)x_1 + \hat{N}_2(\xi, \eta)x_2 + \hat{N}_3(\xi, \eta)x_3 \\ y &= \hat{N}_1(\xi, \eta)y_1 + \hat{N}_2(\xi, \eta)y_2 + \hat{N}_3(\xi, \eta)y_3 \end{aligned} \tag{8.21}$$

It may be verified that the mapping does perform as required. The point $\hat{P}_1$ in the (ξ, η) plane has coordinates $\xi = 0, \eta = 0$ and the shape functions take the values $\hat{N}_1 = 1$ and $\hat{N}_2 = \hat{N}_3 = 0$, so (8.21) gives

$$x = x_1 \quad \text{and} \quad y = y_1$$

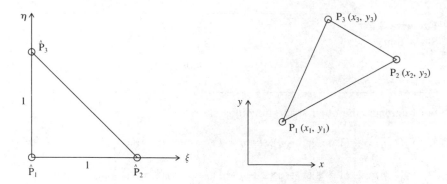

Figure 8.15   The master triangle mapped onto a general straight-sided triangle

That is, the point $\hat{P}_1(0, 0)$ maps to $P_1(x_1, y_1)$. And similarly, $\hat{P}_2$ and $\hat{P}_3$ map to P_2 and P_3, respectively.

To see how the sides of $\hat{P}_1\hat{P}_2\hat{P}_3$ are mapped, consider for example the side $\hat{P}_1\hat{P}_2$, i.e. $\eta = 0$, $0 \le \xi \le 1$. Since

$$\hat{N}_1 = 1 - \xi - \eta, \quad \hat{N}_2 = \xi, \quad \hat{N}_3 = \eta$$

then

$$T_e: \quad x = (1 - \xi)x_1 + \eta x_2 = x_1 + \xi(x_2 - x_1)$$
$$y = (1 - \xi)y_1 + \eta y_2 = y_1 + \xi(y_2 - y_1)$$

If ξ is eliminated, the relation between x and y is

$$y - y_1 = \frac{y_2 - y_1}{x_2 - x_1}(x - x_1)$$

which is clearly the line through P_1 and P_2.

Exercise 8.14
Show that the line $\hat{P}_1\hat{P}_3$ maps to the line P_1P_3.

Exercise 8.15
Form explicitly the mapping T_e for the triangle $P_1(0, 0)$, $P_2(2, 0)$, $P_3(1, 1)$. Verify that the centroid of the triangle $\hat{P}_1\hat{P}_2\hat{P}_3$ maps to the centroid of triangle $P_1P_2P_3$.

8.5.2 Six-noded master element mapped onto a curved-sided triangle

The mapping is again made up of two interpolating formulae,

$$T_e: \quad x = \sum_{i=1}^{6} \hat{N}_i(\xi, \eta)x_i$$

$$y = \sum_{i=1}^{6} \hat{N}_i(\xi, \eta)y_i \qquad (8.22)$$

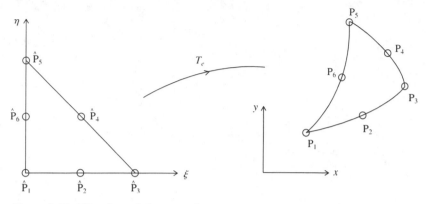

Figure 8.16 The six-noded master element mapped onto a curved-sided triangle

where the shape functions now are the quadratics relevant to this master element obtained in section 8.4.2.

The straight sides of the master element map into curves (which are conics) through the corresponding three points defining the side of the actual element (figure 8.16). The derivation is similar to that which will be given for the eight-noded quadrilateral in the next section.

As a particular degenerate case, it is interesting to note that if the sides of the actual triangle are in fact straight, and if the non-vertex nodes are at the mid-side points, then the mapping does simplify to the linear form of the three-noded triangle. This is shown in the next exercises.

Exercise 8.16
Show that the quadratic shape function sum

$$\hat{N}_1 + \frac{1}{2}(\hat{N}_2 + \hat{N}_6)$$

when simplified, reduces to the linear shape function $\hat{N}_1$ of section 8.4.1. There are two other similar results. Write these down.

Exercise 8.17
Hence show that if P_2, P_4 and P_6 are the mid-points of the sides, i.e.

$$x_2 = \frac{1}{2}(x_1 + x_3) \quad \text{and} \quad y_2 = \frac{1}{2}(y_1 + y_3), \quad \text{etc.}$$

then the mapping becomes linear.

8.5.3 Eight-noded square mapped onto a curved-sided quadrilateral

$\hat{P}_1(-1, -1), \ \hat{P}_2(0, -1), \ldots, \hat{P}_8(-1, 0)$ are to be mapped to $P_1(x_1, y_1)$, $P_2(x_2, y_2), \ldots, P_8(x_8, y_8)$. See figure 8.17.

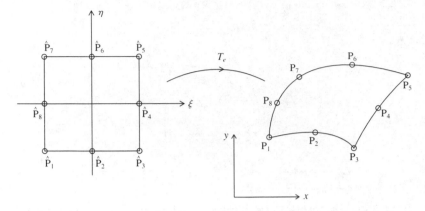

Figure 8.17   The eight-noded square mapped onto a curved-sided quadrilateral

The mapping has the same form as the previous two examples,

$$T_e: \quad x = \sum_{i=1}^{8} \hat{N}_i(\xi, \eta)x_i$$

$$(8.23)$$

$$y = \sum_{i=1}^{8} \hat{N}_i(\xi, \eta)y_i$$

To justify this, if the coordinates of $\hat{P}_1$ are substituted for ξ, η, then

$$\hat{N}_1 = 1 \quad \text{and} \quad \hat{N}_i = 0 \qquad i = 2, 3, \dots, 8.$$

So x becomes x_1 and y becomes y_1 and $(-1, -1)$ maps to P_1. Similarly, $\hat{P}_2$, $\hat{P}_3$, ... map to P_2, P_3, ...

What happens to, say, the line $\hat{P}_1\hat{P}_2\hat{P}_3$ under the mapping? If we set $\eta = -1$ in the shape functions given in (8.18), (8.19) and (8.20),

$$\hat{N}_1 = -\frac{1}{4}(1 - \xi)2\xi \qquad \hat{N}_2 = \frac{1}{2}(1 - \xi^2)(2)$$

$$\hat{N}_3 = \frac{1}{4}(1 + \xi)2\xi \qquad \hat{N}_4 = \hat{N}_5 = \hat{N}_6 = \hat{N}_7 = \hat{N}_8 = 0$$

Thus

$$T_e: \quad x = -\frac{1}{2}\xi(1 - \xi)x_1 + (1 - \xi^2)x_2 + \frac{1}{2}\xi(1 + \xi)x_3$$

$$y = -\frac{1}{2}\xi(1 - \xi)y_1 + (1 - \xi^2)y_2 + \frac{1}{2}\xi(1 + \xi)y_3$$

This is the parametric equation of a curve which passes through P_1, P_2 and P_3. It can be shown that the curve is a conic and so will have no ripples (figure 8.18).

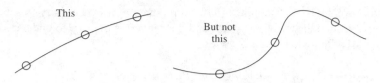

Figure 8.18 Mapping of a side of the master quadrilateral

It will probably not match exactly a given curved boundary of an element, but hopefully it will be close.

8.5.4 The general form of an element mapping

From these examples, it will have become clear that a master element having M nodes and M shape functions $\hat{N}_i, i = 1, 2, \ldots, M$, can be mapped onto an element in the problem domain defined by M corresponding points $P_i(x_i, y_i), i = 1, 2, \ldots, M$, by means of

$$T_e: \quad x = \sum_{i=1}^{M} \hat{N}_i(\xi, \eta) x_i$$

$$y = \sum_{i=1}^{M} \hat{N}_i(\xi, \eta) y_i$$

(8.24)

We now come to the use of the mappings in two-dimensional computation. The work mirrors the earlier illustration by a one-dimension example, but the level of complexity has increased considerably through adding a dimension.

8.6 Use of mappings in two-dimensional element computations

The aim of this section is to express the integral terms in the stiffness matrix and force vector in a computable form. It is hardly practicable to do the arithmetic involved by hand; what follows gives the theory which forms the basis of a computer program to do it for you, if the need ever arises.

To focus our thinking, consider a sample term from a stiffness matrix which has to be re-expressed in (ξ, η) coordinates,

$$\iint_{R^e} \frac{\partial N_i(x, y)}{\partial x} \frac{\partial N_j(x, y)}{\partial x} + \frac{\partial N_i(x, y)}{\partial y} \frac{\partial N_j(x, y)}{\partial y} \, dx \, dy,$$

(8.25)

or a more general term,

$$\iint_{R^e} F\left(\frac{\partial N(x, y)}{\partial x}, \frac{\partial N(x, y)}{\partial y}, N(x, y) \right) dx \, dy$$

(8.26)

where the shape functions N can be linear, quadratic or higher order.

Before we apply the mappings to the computation, it is necessary to set out a little background theory of mappings and their inverses in order to understand the method used and what could go wrong.

8.6.1 A general look at mappings and their inverses

Consider a general mapping from the (ξ, η) plane to the (x, y) plane

$$T: \quad \begin{aligned} x &= x(\xi, \eta) \\ y &= y(\xi, \eta) \end{aligned} \tag{8.27}$$

The direction of the mapping follows the way it is formed in (8.24), from the (ξ, η) plane to the (x, y) plane. However, we need the mapping to exist in the opposite direction as well, because the terms of the stiffness matrix are originally formulated on the actual element using (x, y) coordinates. If the mapping is linear and the actual element has non-zero area, then the inverse will exist; but there are examples of quadratic elements where the 'mid-point' node is so far from the mid-point that the mapped image of the master element overlaps itself and a point in the (x, y) plane comes from more than one point. However, in what follows we will assume that the inverse mapping does exist.

$$T^{-1}: \quad \begin{aligned} \xi &= \xi(x, y) \\ \eta &= \eta(x, y) \end{aligned} \tag{8.28}$$

8.6.2 The Jacobian and derivatives

In transferring terms in the stiffness matrix and force vector such as (8.25) to integrals on the master element, the **Jacobian matrix** plays a part. Some relevant results can be derived from considering the mapping of two nearby points and their return under the inverse mapping.

Suppose that under T, $\hat{P}_0(\xi_0, \eta_0)$ maps to $P_0(x_0, y_0)$ and $\hat{P}(\xi, \eta)$ maps to $P(x, y)$, where $\xi = \xi_0 + \delta\xi$, $\eta = \eta_0 + \delta\eta$, and $x = x_0 + \delta x$, $y = y_0 + \delta y$ (see figure 8.19). Then if $\delta\xi$ and $\delta\eta$ are small, using the chain rule and (8.27)

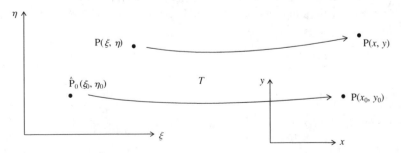

Figure 8.19   Mapping of nearby points

$$\delta x \approx \frac{\partial x}{\partial \xi} \delta \xi + \frac{\partial x}{\partial \eta} \delta \eta$$

$$\delta y \approx \frac{\partial y}{\partial \xi} \delta \xi + \frac{\partial y}{\partial \eta} \delta \eta$$

or in matrix form

$$\begin{bmatrix} \delta x \\ \delta y \end{bmatrix} \approx \begin{bmatrix} \frac{\partial x}{\partial \xi} & \frac{\partial x}{\partial \eta} \\ \frac{\partial y}{\partial \xi} & \frac{\partial y}{\partial \eta} \end{bmatrix} \begin{bmatrix} \delta \xi \\ \delta \eta \end{bmatrix} \tag{8.29}$$

The matrix in this equation

$$\mathbf{J} = \begin{bmatrix} \frac{\partial x}{\partial \xi} & \frac{\partial x}{\partial \eta} \\ \frac{\partial y}{\partial \xi} & \frac{\partial y}{\partial \eta} \end{bmatrix} \tag{8.30}$$

is called the Jacobian matrix of the transformation, and its determinant $|\mathbf{J}|$ is the plain **Jacobian**. Thus the Jacobian is

$$|\mathbf{J}| = \frac{\partial x}{\partial \xi} \frac{\partial y}{\partial \eta} - \frac{\partial x}{\partial \eta} \frac{\partial y}{\partial \xi}$$

Now, under the inverse mapping T^{-1} the points in the (x, y) plane will map back to the same points in the (ξ, η) plane from which they originated under T. Thus, using (8.28), we have

$$\begin{bmatrix} \delta \xi \\ \delta \eta \end{bmatrix} \approx \begin{bmatrix} \frac{\partial \xi}{\partial x} & \frac{\partial \xi}{\partial y} \\ \frac{\partial \eta}{\partial x} & \frac{\partial \eta}{\partial y} \end{bmatrix} \begin{bmatrix} \delta x \\ \delta y \end{bmatrix} \tag{8.31}$$

If the two equations (8.29) and (8.31) are compared, it becomes clear that since the small changes are the same, the two matrices must be mutually inverse. So

$$\begin{bmatrix} \frac{\partial \xi}{\partial x} & \frac{\partial \xi}{\partial y} \\ \frac{\partial \eta}{\partial x} & \frac{\partial \eta}{\partial y} \end{bmatrix} = \begin{bmatrix} \frac{\partial x}{\partial \xi} & \frac{\partial x}{\partial \eta} \\ \frac{\partial y}{\partial \xi} & \frac{\partial y}{\partial \eta} \end{bmatrix}^{-1} \tag{8.32}$$

On obtaining the inverse of $\mathbf{J}$ and equating corresponding terms in the two matrices, we obtain

$$\frac{\partial \xi}{\partial x} = \frac{1}{|\mathbf{J}|} \frac{\partial y}{\partial \eta} \qquad \frac{\partial \xi}{\partial y} = -\frac{1}{|\mathbf{J}|} \frac{\partial x}{\partial \eta}$$

$$\frac{\partial \eta}{\partial x} = -\frac{1}{|\mathbf{J}|} \frac{\partial y}{\partial \xi} \qquad \frac{\partial \eta}{\partial y} = \frac{1}{|\mathbf{J}|} \frac{\partial x}{\partial \xi} \tag{8.33}$$

These equations are important for our purposes because they enable the derivatives with respect to x and y, which are needed but are not available (since the inverse mapping (8.28) is not explicitly formed), to be expressed in terms which are available (from the mapping 8.24).

8.6.3 The effect of the mapping on areas

Another aspect of the mapping which we will need is the effect on areas. The standard result, given in texts on calculus, is that there is a local magnification factor between corresponding areas,

$$\delta A \approx |\mathbf{J}(\xi, \eta)| \delta \hat{A} \tag{8.34}$$

shown in figure 8.20. A derivation of the result is left as an exercise.

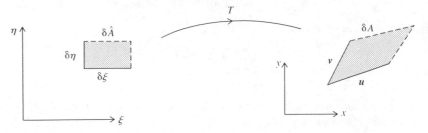

Figure 8.20   Mapping of small areas

Exercise 8.18
Consider the result of mapping a small segment $\delta \xi$ in the (ξ, η) plane. Show that it will be a segment $\mathbf{u}$ in the (x, y) plane given by

$$\mathbf{u} \approx \frac{\partial x}{\partial \xi} \delta \xi \, \mathbf{i} + \frac{\partial y}{\partial \xi} \delta \xi \, \mathbf{j}$$

Form, similarly, the mapping $\mathbf{v}$ of a small segment $\delta \eta$. Then form δA as the area of a parallelogram, $|\mathbf{u} \times \mathbf{v}|$, to obtain the result (8.34).

We now look in more detail at the process of transferring the various items of the integral (8.26) onto the master element.

8.6.4 What happens to shape functions?

Shape functions on the actual element must be replaced by corresponding functions on the master element. For the one-dimensional example at the beginning of the chapter where a linear mapping was used, we showed that the shape function $N_i(x)$ was equivalent to $\hat{N}_i(\xi)$ in the sense that $\hat{N}_i(\xi) = N_i(x(\xi))$, where $x(\xi)$ is the function that defines the mapping. More generally, even if the mapping is non-linear and if the work is in higher dimensions, this approach may be used to form shape functions on the actual element. Thus for a two-dimensional mapping $x = x(\xi, \eta), y = y(\xi, \eta)$ the element shape functions are constructed by

$$N(x, y) = \hat{N}(\xi(x, y), \eta(x, y))$$

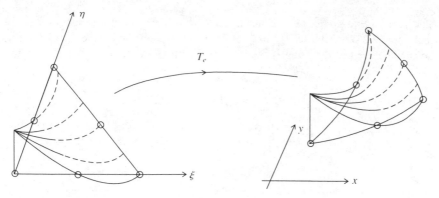

Figure 8.21   The mapping of the shape function $\hat{N}_1(\xi, \eta)$

Geometrically, this says that the actual element shape function takes the same values as the master element shape function at corresponding (mapped) points. To take a particular example, consider a six-noded triangle and N_1. Since $\hat{N}_1 = 1$ at $\hat{P}_1$ and is zero on the lines $\hat{P}_2\hat{P}_6$, and $\hat{P}_3\hat{P}_4\hat{P}_5$, then the mapped function N_1 will be 1 at P_1 and be zero on P_2P_3 and $P_3P_4P_5$. So it will appear as in figure 8.21.

It is tempting to assume that the function $N(x, y)$ derived in this way is a polynomial of the same degree as $\hat{N}(\xi, \eta)$. This is true only if the mapping is linear (if the sides of the actual element are straight and P_2, P_4, P_6 are midpoints). In general $N(x, y)$ will not even be a polynomial, although using these shape functions leads to a method with well-established properties of convergence, as the size of elements is reduced.

Exercise 8.19

Consider whether the following property for polynomial shape functions also holds for the shape functions formed in this new way:

$$\sum_{i=1}^{6} N_i(x, y) = 1.$$

8.6.5 *What happens to differentiation with respect to x and y?*

From

$$N_i(x, y) \quad = \quad \hat{N}_i(\xi, \eta)$$

it follows from using the chain rule that

$$\frac{\partial N_i(x, y)}{\partial x} = \frac{\partial \hat{N}_i(\xi, \eta)}{\partial \xi}\frac{\partial \xi}{\partial x} + \frac{\partial \hat{N}_i(\xi, \eta)}{\partial \eta}\frac{\partial \eta}{\partial x} \tag{8.35}$$

The terms $\partial \hat{N}_i(\xi, \eta)/\partial \xi$ present no difficulty but, because of the direction of the mapping (8.24), the terms $\frac{\partial \xi}{\partial x}$ and $\frac{\partial \eta}{\partial x}$ are not immediately available. They have to be obtained indirectly using (8.33) together with (8.24). Giving, for example,

$$\frac{\partial \xi}{\partial x} = \frac{1}{|\mathbf{J}(\xi, \eta)|} \frac{\partial}{\partial \eta} \left[\sum_{i=1}^{M} \hat{N}_i(\xi, \eta) y_i \right]$$

$$\frac{\partial \xi}{\partial y} = -\frac{1}{|\mathbf{J}(\xi, \eta)|} \frac{\partial}{\partial \eta} \left[\sum_{i=1}^{M} \hat{N}_i(\xi, \eta) x_i \right]$$

Exercise 8.20

Form the two relations for the other derivatives in terms of ξ and η,

$$\frac{\partial \eta}{\partial x} \quad \text{and} \quad \frac{\partial \eta}{\partial y}.$$

8.6.6 What happens to elements of area 'dx dy'?

In two dimensions the magnification factor is the Jacobian (8.34), so

$$dx\, dy = |\mathbf{J}(\xi, \eta)| d\xi\, d\eta \qquad (8.36)$$

8.6.7 The new integral

The original integral had the general form

$$\iint_{R^e} F\left(\frac{\partial N(x, y)}{\partial x}, \frac{\partial N(x, y)}{\partial y}, N(x, y) \right) dx\, dy \qquad (8.37)$$

After the shape functions and their derivatives have been expressed in (ξ, η) coordinates, using sections 8.6.4 and 8.6.5, the integrand F is now given in terms of the new variables and becomes, say $G(\xi, \eta)$. The integral is

$$\iint_{\hat{R}^e} G(\xi, \eta) |\mathbf{J}(\xi, \eta)| \, d\xi\, d\eta \qquad (8.38)$$

so that applying a Gaussian quadrature rule, section 8.3, on the master element $\hat{R}^e$, we have the approximate equivalent

$$\sum_{i=1}^{ngauss} [w_i\, G(\xi_i, \eta_i) |\mathbf{J}(\xi_i, \eta_i)|] \qquad (8.39)$$

where (ξ_i, η_i), w_i, $i = 1, 2 \ldots$, ngauss, are the Gauss points and corresponding weights.

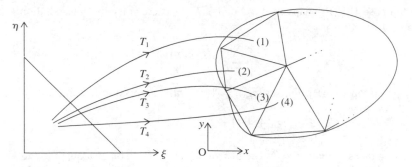

Figure 8.22   Successive mappings

8.7 A look at the combined effect of individual mappings

The above method has been developed for individual elements. The master element is mapped to an actual element and then that element's contribution to the global set of equations is computed and added in. So the whole solution process involves a succession of these mappings: the master element is related to each element in turn, using (8.24), and then the computation is carried out as described in section 8.6. This is illustrated in figure 8.22. The important question is whether this process gives rise to continuity, because on this depends the accuracy and convergence of the solution. Continuity here refers to both the geometry – are there gaps between elements? – and in the problem variable representation – do the neighbouring elements give the same values on the common boundaries? The linear case fairly obviously leads to continuity in both senses, as mentioned in section 5.2, but higher order elements need more detailed examination. Consider two six-noded triangular elements sharing the boundary ABC, as shown in figure 8.23.

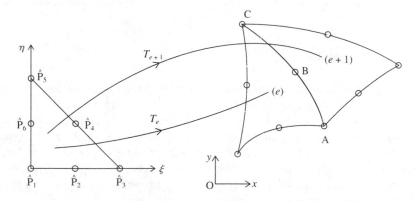

Figure 8.23   Mappings of elements with a common boundary

Suppose $\hat{P}_3\hat{P}_4\hat{P}_5$ maps to ABC under T_e, and consider the interpolation for a general function g:

$$\sum_{i=1}^{6} g_i \hat{N}_i(\xi, \eta)$$

for points on $\hat{P}_3\hat{P}_4\hat{P}_5$. On setting $\eta = 1 - \xi$, the expression becomes

$$2\xi(\xi - \frac{1}{2})g_A + 4\xi(1 - \xi)g_B + 2(1 - \xi)(\frac{1}{2} - \xi)g_C$$

where g takes the value g_A at $\hat{P}_3$, and g_B, g_C at $\hat{P}_4$, $\hat{P}_5$. This is a quadratic in ξ and as a point on the master element moves from $\hat{P}_3$ to $\hat{P}_5$, $1 \geq \xi \geq 0$, then the expression ranges as a (unique) quadratic through g_A, g_B, g_C.

An examination of what happens to the same side ABC under the mapping T_{e+1} gives rise to a similar result, the details of which are given in the following exercise. Since a quadratic is uniquely defined by three points, the two mappings give the same values for g. This applies to the problem variable where g is u, and to the mapping of the geometry, where g is first the x-coordinate and then the y-coordinate. The argument can be developed generally to show that the employment of elements of the same degree leads to a global representation of the geometry, without any gaps or overlaps, and a continuous trial function approximation for the problem variable.

Exercise 8.21
Suppose that under T_{e+1} the side $\hat{P}_1\hat{P}_6\hat{P}_5$ maps to ABC. Show that

$$\sum_{i=1}^{6} g_i \hat{N}_i(\xi, \eta)$$

for points on $\hat{P}_1\hat{P}_6\hat{P}_5$ where $\xi = 0$ becomes a quadratic in the parameter η. Also, that as a point on the master element moves from $\hat{P}_3$ to $\hat{P}_5$, $0 \leq \eta \leq 1$, then the expression ranges as a quadratic through g_A, g_B, g_C.

8.8 Mesh generation

Imagine the difficulty of trying to generate a finite element discretisation of a region such as that shown in figure 8.24. How are the position and type of elements to be chosen? And when this has been completed, the elements and their nodes have to be numbered and the coordinates of each node listed. A big task! Finite element mesh generation has become an area of study of some importance and several approaches have been (and are being) developed. One technique which is useful (though not necessarily the most useful), comes from the idea of mapping from a master element, which has been the theme of this chapter.

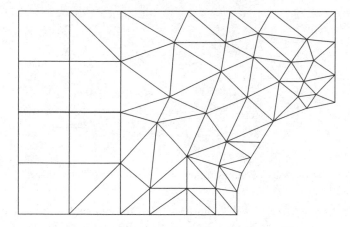

Figure 8.24 A finite element mesh

8.8.1 Mesh generation by mappings

Usually the problem domain is complex but can be broken down into a number of subregions which themselves are fairly simple and can be mapped from a master element with acceptable accuracy. We will consider forming a mesh for a single subregion and then indicate how several such mappings may be combined to form a mesh for a more complicated shape.

A useful master element for this purpose is the eight-noded square which can be mapped into a figure having four curved sides (figure 8.25). Suppose we wish to form a mesh of triangular elements for such a quadrilateral. A mapping T is formed to link the given region R with the master element, using the four vertices

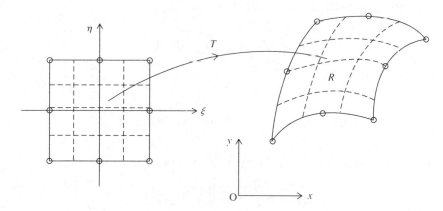

Figure 8.25 Forming a mesh for a single region

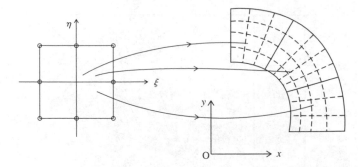

Figure 8.26   Forming a mesh for a more complicated shape

of R and four 'mid-side' nodes chosen so that the mapped sides of the master element fall close to the curved sides of R. If the 'mid-side' nodes are moved towards a vertex, the mesh lines will be concentrated in that area.

A suitable size of mesh must be decided on, and this is specified by the grid lines to be used, rows and columns, in the (ξ, η) plane. These points are evenly spaced and so their coordinates may easily be calculated by a computer program.

The points are then mapped by T to points on R and an element structure may be formed in the (x, y) plane. For example, this may be done by joining straight lines to form straight-sided quadrilaterals which may each then be subdivided into two triangles. It is also possible to form curved elements. A mesh for a more complicated region may be constructed by dividing it into more regularly shaped subregions, forming a mesh for these separately, and then merging the result (figure 8.26).

General exercises for chapter 8

1. Consider forming a two-point Gaussian formula, which contains four arbitrary constants w_1, w_2, x_1, x_2 (figure 8.27):

$$\int_{-1}^{1} f(x)\,dx \approx \sum_{i=1}^{2} w_i f(x_i)$$

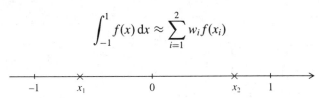

Figure 8.27   Positions for a two-point Gaussian formula

Define an error function for the formula used with a given function f.

$$e(f(x)) = \int_{-1}^{1} f(x)\,dx - \sum_{i=1}^{2} w_i f(x_i).$$

Show that e is linear in f, i.e.

$$e(f_1 + f_2) = e(f_1) + e(f_2) \quad \text{and} \quad e(\alpha f) = \alpha e(f).$$

Deduce that if

$$e(f(x)) = 0 \quad \text{for} \quad f(x) = 1, x, x^2, x^3 \tag{8.40}$$

then it will be zero for $f(x) = a_0 + a_1 x + a_2 x^2 + a_3 x^3$, i.e. any cubic. Now find w_1, w_2, x_1, x_2 so that the four conditions of (8.40) are satisfied. [Note: It is helpful to assume symmetry, i.e. $w_1 = w_2$, and $x_1 = -x_2$.]

2. Show that if a mapping in two dimensions is linear, then straight lines map into straight lines.

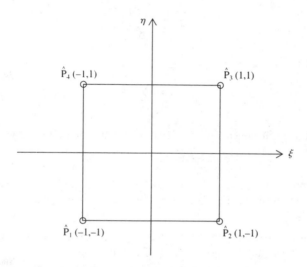

Figure 8.28   Four-noded master quadrilateral

3. Using the shape functions for the four-noded quadrilateral in figure 8.28, form the mapping T which maps $\hat{P}_i, i = 1, 2, 3, 4$, onto the four points $P_i, i = 1, 2, 3, 4$, where

$$P_1(0, 0), \quad P_2(1, 0.5), \quad P_3(1, 1), \quad P_4(0, 1).$$

Show that only the lines parallel to the axes in the (ξ, η) plane are mapped onto straight lines by T.

4. Show that for a linear mapping of the master triangle onto a general three-noded triangle, the centroid of the master triangle maps onto the centroid of the general triangle. Would this be true if the mapping were quadratic? If you think not, look for a counter-example.

5. Consider the mapping of the four-noded square of section 8.2.3 onto the straight-sided quadrilateral with vertices at $(0, 0), (1, 0), (t, t), (0, 1)$, where t is a parameter to be varied. Show that the Jacobian is

$$|\mathbf{J}| = \frac{1}{8}[(1 + 2t) - (1 - t)(\xi + \eta)].$$

Show the following, sketching the line $|\mathbf{J}| = 0$ on the master element in each case, and note the shape of the actual element.

(a) That if $t = 1$, $|\mathbf{J}|$ is never zero, i.e. the mapping is always invertible.

(b) That if $t = \frac{1}{4}$, the Jacobian is positive for all points on the master element.

(c) That if $t = \frac{1}{8}$, then for some points the Jacobian is negative.

6. The mapping

$$T_e: \quad x = \sum_{i=1}^{8} \hat{N}_i(\xi, \eta)x_i$$

$$y = \sum_{i=1}^{8} \hat{N}_i(\xi, \eta)y_i$$

maps the eight-noded master element $\hat{R}^e$ onto the eight nodes

$$P_1(x_1, y_1), P_2(x_2, y_2), \dots, P \dots (x \dots, y \dots).$$

If the nodes P_2, P_4 and P_6 are exactly mid-side nodes, will the mapping T_e degenerate from quadratic to linear ? Either prove the result, or find a counter-example.

7. Show that under a linear invertible mapping, $N_i(x, y)$ becomes $\hat{N}_i(\xi, \eta)$, where N is a shape function polynomial of any degree.

8. Write down the mapping of the master element $(0, 0), (1, 0)$ and $(0, 1)$ onto the general straight-sided triangle joining

$$P_i(x_i, y_i), \quad P_j(x_j, y_j) \quad \text{and} \quad P_k(x_k, y_k).$$

Using the theory given in section 8.6, show that a term in the element stiffness matrix for Laplace's equation is given by

$$K_{ij}^e = \frac{1}{2|\mathbf{J}|}[a_i a_j + b_i b_j]$$

where $a_i = x_j - x_k$, $b_i = y_j - y_k$, etc., and $|\mathbf{J}| = a_j b_k - a_k b_j$. Write down the complete element stiffness matrix.

9. Consider whether there is continuity on a common boundary AB between a four-noded quadrilateral and a three-noded triangle (figure 8.29).

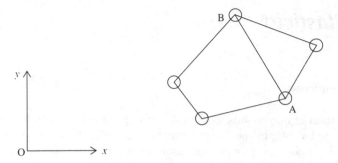

Figure 8.29   A four-noded quadrilateral joining a three-noded triangle

10. Consider whether there is continuity across the common boundary ABC between the six-noded triangle and two three-noded triangles in figure 8.30.

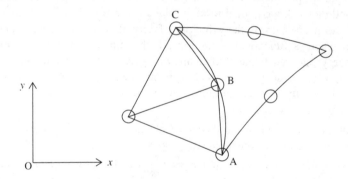

Figure 8.30   A six-noded triangle joining two three-noded triangles

9 Elasticity

9.1 Introduction

The intention of this book is, besides describing the theory of finite elements, to illustrate it by simple numerical examples in order to increase the reader's confidence. Elasticity provides a more severe test, for it is a complicated phenomenon. However, it is a very important field for the use of finite elements; it was in this setting that it was born, in the need to analyse stresses in aircraft, and it has remained probably the most important application ever since. So, although this may be the most difficult chapter, the importance of the topic dictates that it should be included.

Elasticity is described mathematically by a system of partial differential equations, which compares with the single second-order equation that models field (Laplace) problems. The passage from a system of equations to the corresponding weak variational form involves a fair amount of mathematics. The Galerkin form ends up as the principle of virtual work and, in the alternative approach where a functional is minimised, that functional turns out to be the strain energy integral. If the mathematics involved is not of interest, the reader may wish to skip section 9.3 and start at section 9.4, assuming the well-established (by other means) principles in the theory of elasticity.

This chapter:

- Outlines sufficient of the theory of elasticity in order to present the finite element method in this context.

- Charts the development of the system of elasticity equations into weak variational form.

- Applies the finite element method with linear elements and obtains a general formula for the stiffness matrix and force vector.

- Works through a simple numerical example, which sadly is not so simple numerically.

9.2 Background to elasticity

Elasticity is the study of the deformation of a solid body under loading, together with the resulting stresses and strains. It is assumed that the body will recover its original size and shape after the forces causing the change are removed. A general discussion of elasticity is a task not to be undertaken lightly; our ambition in this section is to present sufficient of the theory of two-dimensional elasticity so that the reader may understand this application of the finite element technique.

166

Figure 9.1   A thin body suitable for analysis as plane stress

Gravitational forces will be considered, but not the stress resulting from temperature differences.

9.2.1 *Plane stress and plane strain*

Elasticity is essentially a three-dimensional phenomenon, but in two cases it may be analysed by working in only two dimensions.

Plane stress
For a plate, i.e. a thin plane body (figure 9.1), loaded only in the plane of the plate, the stress components σ_{zz}, σ_{zx} and σ_{zy} are all zero on both faces and small within the plate. The assumption for 'plane stress' is that they may be ignored through the thickness.

As a consequence of this assumption, the shear strains ϵ_{zx} and ϵ_{zy} are zero, but the normal strain ϵ_{zz} is not, because the in-plane stresses cause deformation normal to the plane of the plate (an effect quantified by Poisson's ratio).

Plane strain
This occurs when a body is free to expand or to contract only in the plane in which the loads are applied. For example, a long cylinder with the same loading all along its length (figure 9.2) may be analysed by considering a cross-section in plane strain.

In the two cases the mathematical analysis is very similar, the difference being in a matrix **D** (to be mentioned later) which is used in defining Hooke's law. Also, the thickness of the body may be left out of the consideration; in plane strain the body may be thought of as having unit length, and for plane stress the thickness t will be seen to be a common factor which may be divided out.

9.2.2 *Displacement and strains*

The example in chapter 2, the extension of an elastic string, illustrates strain in one dimension.

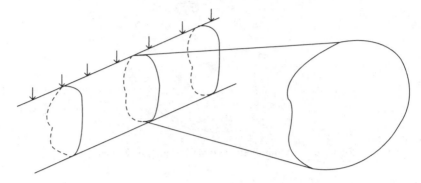

Figure 9.2   A long cylinder suitable for analysis as plane strain

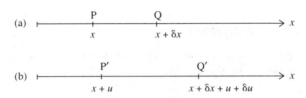

Figure 9.3   A section of an elastic string (a) before, (b) after, stretching

As a result of an applied force, a segment PQ deforms to P′Q′ (figure 9.3). Clearly, the increase in length, δu, depends on the original length, δx. The strain is defined as the increase per unit length,

$$\epsilon_x = \lim_{\delta x \to 0} \frac{\delta u}{\delta x} = \frac{du}{dx}$$

In two dimensions, the basic problem variable is a vector **u** made up of two components, u and v, the displacements in the x- and y-directions (figure 9.4).

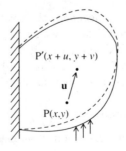

Figure 9.4   A body before and after (dotted) displacement

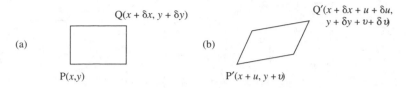

Figure 9.5   A small rectangle (a) before, (b) after distortion

The concept of strain naturally extends to the two coordinate directions, but in addition there is also a shearing strain. The distortion of a small rectangle set in the body as a result of an applied force is shown in figure 9.5, which corresponds to figure 9.3 for the one-dimensional case.

If the original rectangle were distorted into a similar rectangle, then the only direct strains

$$\epsilon_{xx} = \frac{\partial u}{\partial x} \qquad \epsilon_{yy} = \frac{\partial v}{\partial y}$$

would arise. However, there is a more general distortion into the four-sided (nearly a parallelogram) figure 9.5(b). This complication is quantified by the shear strain

$$\gamma_{xy} = \frac{\partial u}{\partial y} + \frac{\partial v}{\partial x}$$

Thus there are three strain components which are given in terms of the displacement vector by

$$\boldsymbol{\epsilon} = \begin{bmatrix} \epsilon_{xx} \\ \epsilon_{yy} \\ \gamma_{xy} \end{bmatrix} = \begin{bmatrix} \frac{\partial}{\partial x} & 0 \\ 0 & \frac{\partial}{\partial y} \\ \frac{\partial}{\partial y} & \frac{\partial}{\partial x} \end{bmatrix} \begin{bmatrix} u \\ v \end{bmatrix} = \mathbf{Su} \tag{9.1}$$

where $\mathbf{S}$ defined by this equation is the matrix operator

$$\mathbf{S} = \begin{bmatrix} \frac{\partial}{\partial x} & 0 \\ 0 & \frac{\partial}{\partial y} \\ \frac{\partial}{\partial y} & \frac{\partial}{\partial x} \end{bmatrix} \tag{9.2}$$

9.2.3 Stresses

Consider a small line segment AB, of length δy, set in the body so that the direction of its normal is the direction of the x-axis (figure 9.6). Let the force $\mathbf{F}$, with components F_x and F_y, represent the action of the body on the right of AB on the left side, i.e. the force that the body on the left experiences. Stress has units of force per unit area; thus if δy is small, so that $\mathbf{F}$ does not vary much over AB, the stress normal to the segment AB is given by

$$\sigma_{xx} \approx \frac{F_x}{t\delta y}$$

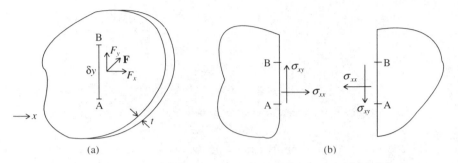

(a)   (b)

Figure 9.6   (a) A line segment with normal in the *x*-direction; (b) stresses on either side of AB

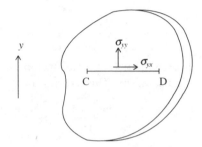

Figure 9.7   A line segment with normal in the *y*-direction

where t is the thickness. This is a direct or normal stress; the other component F_y gives rise to a shearing stress

$$\sigma_{xy} \approx \frac{F_y}{t\delta y}$$

Note that owing to the principle of action and reaction, the forces that the body on the right experiences are equal but opposite to the forces felt by the body on the left. The resulting stresses on opposite sides of segment AB are therefore equal in magnitude, but opposite in direction (see figure 9.6b).

Similarly, a segment CD of length δx having its normal in the *y*-direction will have stresses σ_{yy} and σ_{yx}, as shown in figure 9.7.

9.2.4 Equilibrium equations

Consider the equilibrium of a small rectangle, set in the body, with sides in the directions of the axes. The stresses experienced by the interior of the rectangle are shown in figure 9.8, where allowance has been made for the changes in stress arising from different positions. In addition, we assume that the body is under the action of a distributed force per unit volume, **b**, with components b_x and b_y. A

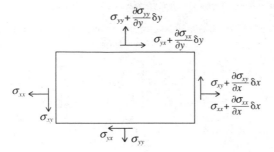

Figure 9.8   Stresses on the sides of a small rectangle

common example of this type of force is the case of self-weight, where $b_x = 0$ and $b_y = -\rho g$, ρ is the density and g the acceleration due to gravity.

Resolving the forces (stress × area) in the x- and y-directions gives

$$\left[\frac{\partial \sigma_{xx}}{\partial x} \delta x\right] t \delta y + \left[\frac{\partial \sigma_{yx}}{\partial y} \delta y\right] t \delta x + t b_x \delta x \delta y = 0 \tag{9.3}$$

$$\left[\frac{\partial \sigma_{yy}}{\partial y} \delta y\right] t \delta x + \left[\frac{\partial \sigma_{xy}}{\partial x} \delta x\right] t \delta y + t b_y \delta x \delta y = 0. \tag{9.4}$$

And taking moments about the centre

$$[\sigma_{xy} t \delta y]\delta x + \left[\frac{\partial \sigma_{xy}}{\partial x} \delta x\right](t\delta y)\frac{\delta x}{2} = [\sigma_{yx} t \delta x]\delta y + \left[\frac{\partial \sigma_{yx}}{\partial y} \delta y\right](t\delta x)\frac{\delta y}{2}$$

i.e.
$$\sigma_{xy} + \frac{1}{2}\frac{\partial \sigma_{xy}}{\partial x}\delta x = \sigma_{yx} + \frac{1}{2}\frac{\partial \sigma_{yx}}{\partial y}\delta y$$

Letting $\delta x, \delta y \to 0$ the two shearing stresses σ_{xy} and σ_{yx} are seen to be equal; then (9.3) and (9.4) become

$$\frac{\partial \sigma_{xx}}{\partial x} + \frac{\partial \sigma_{xy}}{\partial y} + b_x = 0$$

$$\frac{\partial \sigma_{yy}}{\partial y} + \frac{\partial \sigma_{xy}}{\partial x} + b_y = 0.$$

In matrix form

$$\begin{bmatrix} \frac{\partial}{\partial x} & 0 & \frac{\partial}{\partial y} \\ 0 & \frac{\partial}{\partial y} & \frac{\partial}{\partial x} \end{bmatrix}\begin{bmatrix} \sigma_{xx} \\ \sigma_{yy} \\ \sigma_{xy} \end{bmatrix} + \begin{bmatrix} b_x \\ b_y \end{bmatrix} = \mathbf{0} \tag{9.5}$$

or
$$\mathbf{S}^{\mathrm{T}}\boldsymbol{\sigma} + \mathbf{b} = \mathbf{0}$$

where $\mathbf{S}$ is the same operator as in (9.2) and $\boldsymbol{\sigma} = [\,\sigma_{xx}\ \sigma_{yy}\ \sigma_{xy}\,]^{\mathrm{T}}$.

9.2.5 Relation between stress and strain

According to the theory of elasticity, stresses and strains are directly related. This
is expressed by Hooke's law which, for plane stress, is

$$\begin{bmatrix} \sigma_{xx} \\ \sigma_{yy} \\ \sigma_{xy} \end{bmatrix} = \frac{E}{1-v^2} \begin{bmatrix} 1 & v & 0 \\ v & 1 & 0 \\ 0 & 0 & \frac{1-v}{2} \end{bmatrix} \begin{bmatrix} \epsilon_{xx} \\ \epsilon_{yy} \\ \gamma_{xy} \end{bmatrix}$$

i.e. $$\boldsymbol{\sigma} = \mathbf{D}\boldsymbol{\epsilon} \qquad\qquad (9.6)$$

where E is the modulus of elasticity and v is Poisson's ratio.

If a body satisfies the condition for plane strain, the matrix $\mathbf{D}$ in Hooke's law is

$$\mathbf{D} = \frac{E(1-v)}{(1+v)(1-2v)} \begin{bmatrix} 1 & \frac{v}{(1-v)} & 0 \\ \frac{v}{(1-v)} & 1 & 0 \\ 0 & 0 & \frac{1-2v}{2(1-v)} \end{bmatrix}$$

9.2.6 Boundary conditions

The boundary conditions are usually a combination of prescribed displacements –
the body is fixed in a certain way – and prescribed applied forces. Known
displacements are incorporated into the trial function, whereas the applied forces
give rise to a term in the variational formulation. The relation between the
internal stresses and an applied force are considered in this section.

Consider an applied boundary force per unit area, $\mathbf{p}$, with components p_x and
p_y, and the equilibrium of a small triangle with two sides parallel to the axes and
the third side of length δs approximating to the boundary (figure 9.9).

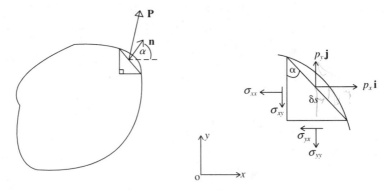

Figure 9.9   A boundary force and in detail

$$p_x t \delta s = \sigma_{xx} t \delta y + \sigma_{yx} t \delta x$$
$$p_y t \delta s = \sigma_{yy} t \delta y + \sigma_{xy} t \delta y$$

or

$$p_x = \sigma_{xx} \frac{\delta y}{\delta s} + \sigma_{yx} \frac{\delta x}{\delta s}$$
$$p_y = \sigma_{yy} \frac{\delta x}{\delta s} + \sigma_{xy} \frac{\delta y}{\delta s}$$

Introducing the angle α between the unit normal $\mathbf{n}$ and the x-axis, and writing

$$n_x = \mathbf{n} \cdot \mathbf{i} = \cos \alpha = \frac{\delta y}{\delta s}, \qquad n_y = \mathbf{n} \cdot \mathbf{j} = \sin \alpha = \frac{\delta x}{\delta s}$$

this becomes

$$p_x = n_x \sigma_{xx} + n_y \sigma_{yx}$$
$$p_y = n_y \sigma_{yy} + n_x \sigma_{xy}.$$

Noting that $\sigma_{xy} = \sigma_{yx}$

$$\begin{bmatrix} p_x \\ p_y \end{bmatrix} = \begin{bmatrix} n_x & 0 & n_y \\ 0 & n_y & n_x \end{bmatrix} \begin{bmatrix} \sigma_{xx} \\ \sigma_{yy} \\ \sigma_{xy} \end{bmatrix}$$

i.e.

$$\mathbf{p} = \mathbf{M}^{\mathrm{T}} \boldsymbol{\sigma} \tag{9.7}$$

Note that

$$\mathbf{M}^{\mathrm{T}} = \begin{bmatrix} n_x & 0 & n_y \\ 0 & n_y & n_x \end{bmatrix}$$

has the same structure as the matrix operator, $\mathbf{S}$, of (9.1) and (9.5), with $\frac{\partial}{\partial x}$ and $\frac{\partial}{\partial y}$ replaced by n_x and n_y, respectively.

9.2.7 Summary

Strain/displacement relation	$\boldsymbol{\epsilon} = \mathbf{S}\mathbf{u}$
Stress/strain relation (Hooke's law)	$\boldsymbol{\sigma} = \mathbf{D}\boldsymbol{\epsilon}$
Equilibrum equation	$\mathbf{S}^{\mathrm{T}}\boldsymbol{\sigma} + \mathbf{b} = \mathbf{0}$
Applied stress boundary conditions	$\mathbf{M}^{\mathrm{T}}\boldsymbol{\sigma} = \mathbf{p}$

9.3 Variational forms for elasticity

The structure of the derivation of the weak variational form of the equations of elasticity is similar to that used with Laplace's equation in chapter 6; however, the details are more complicated. A system of equations (two) is being considered rather than a single equation, which translates into working with matrices rather

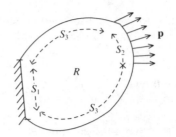

Figure 9.10   Problem description

than scalars. The manipulation concludes with two forms of the weak variational statement of the problem which correspond to the well-known principles of virtual work and minimising strain energy.

9.3.1 Stating the problem

The displacement vector $\mathbf{u}$ is specified on part of the surface S_1(see figure 9.10), and a loading is applied in the form of point forces, distributed surface forces $\mathbf{p}$, or volume forces $\mathbf{b}$. The three equations, $\boldsymbol{\epsilon} = \mathbf{Su}$, $\boldsymbol{\sigma} = \mathbf{D\epsilon}$ and $\mathbf{S}^T\boldsymbol{\sigma} = \mathbf{0}$ may be combined into a system of equations (two) for $\mathbf{u}$,

$$\mathbf{S}^T\mathbf{DSu} + \mathbf{b} = \mathbf{0} \tag{9.8}$$

Note that since $\mathbf{S}$ is a matrix operator in $\frac{\partial}{\partial x}$ and $\frac{\partial}{\partial y}$, the matrix equation (9.8) is a pair of second-order partial differential equations in the components u and v.

Exercise 9.1
Form the product $\mathbf{S}^T\mathbf{DS}$ as a single matrix for the case of plane stress.

Thus $\mathbf{u}$ satisfies equation (9.8) on R, with boundary conditions

$$\mathbf{u} = \mathbf{u}^* \text{ on } S_1$$

$$\mathbf{M}^T\mathbf{DSu} = \begin{cases} \mathbf{p} & \text{on } S_2 \\ \mathbf{0} & \text{on } S_3 \end{cases} \tag{9.9}$$

where $S_1 \cup S_2 \cup S_3$ is the boundary of R.

To change this into variational form we proceed as in chapter 6. Choose a trial function $\mathbf{u}(x, y) = [u(x, y), v(x, y)]^T$(it is convenient, though not very accurate, to use the same symbol $\mathbf{u}$ as in the equation) and form the residual when this is substituted into (9.8)

$$\mathbf{r}(x, y) = \mathbf{S}^T\mathbf{DSu}(x, y) + \mathbf{b}(x, y)$$

The test function $\mathbf{w}(x, y) = [w_1(x, y), w_2(x, y)]^T$ needed to develop the variational form must contain two components w_1 and w_2 which serve the function of zeroing the two components of $\mathbf{r}$.

Using the fundamental lemma, **u** must satisfy: $\mathbf{u} = \mathbf{u}^*$ on S_1 and

$$\iint_R \mathbf{w}^{\mathrm{T}}(\mathbf{S}^{\mathrm{T}}\mathbf{D}\mathbf{S}\mathbf{u} + \mathbf{b})\, \mathrm{d}A = \mathbf{0} \tag{9.10}$$

for all **w** such that $\mathbf{w} = \mathbf{0}$ on S_1.

Note that equation (9.10) can be written in terms of the stress as

$$\iint_R \mathbf{w}^{\mathrm{T}}(\mathbf{S}^{\mathrm{T}}\boldsymbol{\sigma} + \mathbf{b})\, \mathrm{d}A = \mathbf{0} \tag{9.11}$$

Exercise 9.2

Verify that the matrix $\mathbf{w}^{\mathrm{T}}\mathbf{S}^{\mathrm{T}}\mathbf{D}\mathbf{S}\mathbf{u}$ has dimensions 1×1.

In order to obtain the weak variational form, the order (two) of the derivatives acting on **u** has to be reduced by transferring one onto **w**. This is an 'integration by parts' process, the details of which appear in appendices 1 and 2 at the end of the chapter, in order not to disturb the flow of the argument. These two results are stated now in the context of the current problem. The first corresponds to the formula for differentiation of a product.

$$\mathrm{tr}\big[\mathbf{S}^{\mathrm{T}}(\boldsymbol{\sigma}\mathbf{w}^{\mathrm{T}})\big] = \mathbf{w}^{\mathrm{T}}\mathbf{S}^{\mathrm{T}}\boldsymbol{\sigma} + \boldsymbol{\sigma}^{\mathrm{T}}\mathbf{S}\mathbf{w}, \tag{9.12}$$

where 'tr' denotes the **trace** of a matrix, which is the sum of the diagonal terms. For a general $n \times n$ matrix **A**,

$$\mathrm{tr}[\mathbf{A}] = \sum_{i=1}^{n} A_{ii}.$$

The second is a generalisation of the divergence theorem where a domain integral is equated to a boundary integral

$$\iint_R \mathrm{tr}\big[\mathbf{S}^{\mathrm{T}}(\boldsymbol{\sigma}\mathbf{w}^{\mathrm{T}})\big]\, \mathrm{d}A = \int_S \mathrm{tr}\big[\mathbf{M}^{\mathrm{T}}(\boldsymbol{\sigma}\mathbf{w}^{\mathrm{T}})\big]\, \mathrm{d}s$$

$$= \int_S \mathbf{w}^{\mathrm{T}}\mathbf{M}^{\mathrm{T}}\boldsymbol{\sigma}\, \mathrm{d}s \tag{9.13}$$

Exercise 9.3

Verify that
$$\mathrm{tr}\big[\mathbf{M}^{\mathrm{T}}(\boldsymbol{\sigma}\mathbf{w}^{\mathrm{T}})\big] = \mathbf{w}^{\mathrm{T}}\mathbf{M}^{\mathrm{T}}\boldsymbol{\sigma}$$

From (9.11) and (9.12) the problem may be stated as: find **u** satisfying $\mathbf{u} = \mathbf{u}^*$ on S_1 and

$$\iint_R \{\mathrm{tr}\big[\mathbf{S}^{\mathrm{T}}(\boldsymbol{\sigma}\mathbf{w}^{\mathrm{T}})\big] - \boldsymbol{\sigma}^{\mathrm{T}}\mathbf{S}\mathbf{w} + \mathbf{w}^{\mathrm{T}}\mathbf{b}\}\, \mathrm{d}A = 0$$

for all **w** such that $\mathbf{w} = \mathbf{0}$ on S_1.

Using (9.13)

$$\int\int_S \mathbf{w}^T \mathbf{M}^T \boldsymbol{\sigma} \, ds - \int\int_R \boldsymbol{\sigma}^T \mathbf{Sw} \, dA + \int\int_R \mathbf{w}^T \mathbf{b} \, dA = 0$$

The boundary integral can be modified because $\mathbf{w} = \mathbf{0}$ on S_1, and through the boundary condition (9.9). The equation then becomes

$$\int\int_R \boldsymbol{\sigma}^T \mathbf{Sw} \, dA = \int_{S_2} \mathbf{w}^T \mathbf{p} \, ds + \int\int_R \mathbf{w}^T \mathbf{b} \, dA.$$

It may be further modified by writing $\boldsymbol{\epsilon}_w$ for the strain energy corresponding to the test function $\mathbf{w}$, i.e. $\boldsymbol{\epsilon}_w = \mathbf{Sw}$ and noting that $\boldsymbol{\sigma}^T \boldsymbol{\epsilon}_w = \boldsymbol{\epsilon}_w^T \boldsymbol{\sigma}$,

$$\int\int_R \boldsymbol{\epsilon}_w^T \boldsymbol{\sigma} \, dA = \int_{S_2} \mathbf{w}^T \mathbf{p} \, ds + \int\int_R \mathbf{w}^T \mathbf{b} \, dA \tag{9.14}$$

This statement of the weak variational form is the principle of virtual displacements, represented here by the test function $\mathbf{w}$.

If, for any set of virtual displacements, the work done by the external forces is equal to the work done by the internal stresses, then the system is in equilibrium.

The form (9.14) is a particular case of the general form $B(\mathbf{u}, \mathbf{w}) = L(\mathbf{w})$ of chapter 6. Here

$$B(\mathbf{u}, \mathbf{w}) = \int\int_R \boldsymbol{\sigma}^T \mathbf{Sw} \, dA \quad \text{and} \quad L(\mathbf{w}) = \int_{S_2} \mathbf{w}^T \mathbf{p} \, ds + \int\int_R \mathbf{w}^T \mathbf{b} \, dA$$

Exercise 9.4
Show that B is bilinear and symmetric.

In order to introduce the other weak variational form, that of minimising a functional, it is necessary to show that $B(\mathbf{u}, \mathbf{u})$ is positive definite. Now

$$B(\mathbf{u}, \mathbf{u}) = \int\int_R (\mathbf{Su})^T \mathbf{DSu} \, dA$$

$$= \int\int \boldsymbol{\epsilon}^T \mathbf{D}\boldsymbol{\epsilon} \, dA$$

Alternatively

$$B(\mathbf{u}, \mathbf{u}) = \int\int \boldsymbol{\sigma}^T \boldsymbol{\epsilon} \, dA$$

This last integral is twice the elastic strain energy of the body, which has to be positive for real materials.

Thus the equilibrum displacement is also given by $\mathbf{u}$ satisfying $\mathbf{u} = \mathbf{u}^*$ on S_1 and minimising

$$V(\mathbf{u}) = \frac{1}{2} \int\!\!\int_R \boldsymbol{\sigma}^T \boldsymbol{\epsilon} \ dA - \int_{S_2} \mathbf{u}^T \mathbf{p} \ ds - \int\!\!\int_R \mathbf{u}^T \mathbf{b} \ dA \qquad (9.15)$$

The functional may be recognised as the total energy resulting from a displacement $\mathbf{u}$ which includes the elastic strain energy and the potential energy due to the external applied forces $\mathbf{p}$ and $\mathbf{b}$.

9.4 A simple illustrative problem

Consider the distortion of a simply supported thin square block under a uniformly distributed load (figure 9.11). The plane stress model is suitable, and because of symmetry it is necessary to analyse only half of the block. The boundary conditions are as shown; the vertical section through the middle of the block will not move in a horizontal direction as indicated by the constraints. The block may conveniently be discretised into four right-angled isosceles triangles. This is shown in figure 9.12.

There is probably not much extra work involved if, before approaching the computations for this particular problem with right-angled isosceles elements, we look at the general element and establish the structure for its stiffness matrix. So we will proceed in this way.

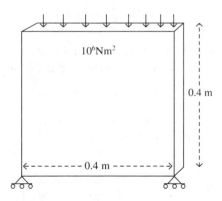

Figure 9.11   A simple problem to illustrate the method

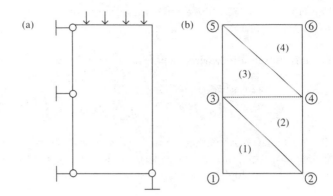

Figure 9.12   (a) The reduced problem and (b) its discretisation

9.5 The general finite element approach and linear triangles

9.5.1 The finite element approach

The finite element trial function is substituted for **u**. Suppose that piecewise linear basis functions (see section 7.4) are used for the two components of **u**, i.e.

$$u = \sum_{i=1}^{N} u_i \phi_i(x, y) \qquad v = \sum_{i=1}^{N} v_i \phi_i(x, y) \tag{9.16}$$

where N is the number of nodes and u_i, v_i are the components of **u** at node i. Equation (9.16) may be rewritten as

$$\mathbf{u} = \sum_{i=1}^{N} \mathbf{u}_i \phi_i(x, y). \tag{9.17}$$

It is more usual to write the N pairs of unknowns (u_i, v_i), $i = 1, 2, \ldots, N$, as a single vector of length $2N$, with the us and vs alternating, i.e.

$$\mathbf{U} = [U_1, U_2, U_3, \ldots, U_{2N}]^{\mathrm{T}} = [u_1, v_1, u_2, v_2, \ldots, u_N, v_N]^{\mathrm{T}}$$

where $U_{2i-1} = u_i$, $U_{2i} = v_i$, for $i = 1, 2, \ldots, N$.

With this notation (9.17) may be written as

$$\mathbf{u} = \begin{bmatrix} \phi_1 & 0 & \phi_2 & 0 & \cdots & 0 \\ 0 & \phi_1 & 0 & \phi_2 & \cdots & \phi_N \end{bmatrix} \begin{bmatrix} U_1 \\ U_2 \\ U_3 \\ \vdots \\ U_{2N} \end{bmatrix}$$

or, briefly

$$\mathbf{u} = \mathbf{\Phi U}$$

Setting this into (9.14), where $\boldsymbol{\epsilon}_w = \mathbf{Sw}$ and $\boldsymbol{\sigma} = \mathbf{DSu}$

$$\int\int_R \boldsymbol{\epsilon}_w^T \mathbf{DS\Phi U}\, dA = \int_{S_2} \mathbf{w}^T \mathbf{p}\, ds + \int\int_R \mathbf{w}^T \mathbf{b}\, dA$$

or

$$\left[\int\int_R \boldsymbol{\epsilon}_w^T \mathbf{DS\Phi}\, dA\right]\mathbf{U} = \int_{S_2} \mathbf{w}^T \mathbf{p}\, ds + \int\int_R \mathbf{w}^T \mathbf{b}\, dA,$$

which is a single equation with the $2N$ unknowns of $\mathbf{U}$. To reduce the residual in the region of node 1 for the first equation of the pair (9.8) corresponding to the displacement u_1, the test function is chosen according to the Galerkin method as

$$\mathbf{w} = [\phi_1, 0]^T$$

and for the second corresponding to v_1 as,

$$\mathbf{w} = [0, \phi_1]^T$$

Continuing this process for all the nodes gives the $2N$ choices for $\mathbf{w}$ necessary to equate the number of equations and unknowns. Thus, the equations are:

$$\left\{\int\int_R \left(\mathbf{S}\begin{bmatrix}\phi_i \\ 0\end{bmatrix}\right)^T \mathbf{DS\Phi}\, dA\right\}\mathbf{U} = \int_{S_2} [\phi_i, 0]\mathbf{p}\, ds + \int\int_R [\phi_i, 0]\mathbf{b}\, ds$$

$$\left\{\int\int_R \left(\mathbf{S}\begin{bmatrix}0 \\ \phi_i\end{bmatrix}\right)^T \mathbf{DS\Phi}\, dA\right\}\mathbf{U} = \int_{S_2} [0, \phi_i]\mathbf{p}\, ds + \int\int_R [0, \phi_i]\mathbf{b}\, ds$$

where both equations hold for $i = 1, 2, \ldots, N$.

These may be combined into a single matrix equation

$$\left\{\int\int_R (\mathbf{S\Phi})^T \mathbf{DS\Phi}\, dA\right\}\mathbf{U} = \int_{S_2} \mathbf{\Phi}^T \mathbf{p}\, ds \tag{9.18}$$

Exercise 9.5
Verify that

$$(\mathbf{S\Phi})^T \mathbf{DS\Phi}$$

is a $2N \times 2N$ matrix.

The integral over R is now replaced by a sum of integrals over the elements

$$\int\int_R * \, dA = \sum_{e=1}^{E} \int\int_{R^e} * \, dA \tag{9.19}$$

i.e.

$$\sum_{e=1}^{E}\left\{\int\int_{R^e}(\mathbf{S\Phi})^{\mathrm{T}}\mathbf{DS\Phi}\,\mathrm{d}A\right\}\mathbf{U} = \sum_{e=1}^{E}\left\{\int_{S^e}\mathbf{\Phi}^{\mathrm{T}}\mathbf{p}\,\mathrm{d}s + \int\int_{R^e}\mathbf{\Phi}^{\mathrm{T}}\mathbf{b}\,\mathrm{d}A\right\}$$

or

$$\left\{\sum_{e=1}^{E}\mathbf{K}^e\right\}\mathbf{U} = \sum_{e=1}^{E}\mathbf{f}^e$$

or, briefly

$$\mathbf{KU} = \mathbf{f}$$

where

$$\mathbf{K} = \sum_{e=1}^{E}\mathbf{K}^e \quad \text{and} \quad \mathbf{f} = \sum_{e=1}^{E}\mathbf{f}^e.$$

$$\mathbf{K}^e = \int\int_{R^e}(\mathbf{S\Phi})^{\mathrm{T}}\mathbf{DS\Phi}\,\mathrm{d}A \quad \text{and} \quad \mathbf{f}^e = \int_{S^e}\mathbf{\Phi}^{\mathrm{T}}\mathbf{p}\,\mathrm{d}s + \int\int_{R^e}\mathbf{\Phi}^{\mathrm{T}}\mathbf{b}\,\mathrm{d}A \quad (9.20)$$

We will now consider in detail the element stiffness matrix $\mathbf{K}^e$, and later the force vector $\mathbf{f}^e$.

9.5.2 The stiffness matrix for a general linear triangle

Only ϕ_i, ϕ_j and ϕ_k are non-zero on element e, so we follow the approach of section 4.2 and calculate the condensed form of the element stiffness matrix and force vector for the triangle shown in figure 9.13.

Now

$$\mathbf{\Phi}^e = \begin{bmatrix} \phi_i & 0 & \phi_j & 0 & \phi_k & 0 \\ 0 & \phi_i & 0 & \phi_j & 0 & \phi_k \end{bmatrix}$$

$$= \begin{bmatrix} N_i & 0 & N_j & 0 & N_k & 0 \\ 0 & N_i & 0 & N_j & 0 & N_k \end{bmatrix}$$

where the basis functions ϕ have been replaced by the corresponding shape functions N on e, and

$$\mathbf{U}^e = \begin{bmatrix} U_{2i-1}, U_{2i}, U_{2j-1}, U_{2j}, U_{2k-1}, U_{2k} \end{bmatrix}^{\mathrm{T}} = [u_i, v_i, u_j, v_j, u_k, v_k]^{\mathrm{T}}$$

As part of the process of forming $\mathbf{K}_{\mathrm{c}}^e$ consider the estimated strains obtained from the linear finite element assumption

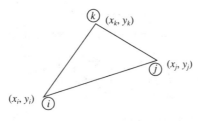

Figure 9.13   The general linear triangle

$$\boldsymbol{\epsilon} = \mathbf{S}\boldsymbol{\phi}^e\mathbf{U}^e = \begin{bmatrix} \frac{\partial}{\partial x} & 0 \\ 0 & \frac{\partial}{\partial y} \\ \frac{\partial}{\partial y} & \frac{\partial}{\partial x} \end{bmatrix} \begin{bmatrix} N_i & 0 & N_j & 0 & N_k & 0 \\ 0 & N_i & 0 & N_j & 0 & N_k \end{bmatrix} \mathbf{U}^e$$

$$= \begin{bmatrix} \frac{\partial N_i}{\partial x} & 0 & \frac{\partial N_j}{\partial x} & 0 & \frac{\partial N_k}{\partial x} & 0 \\ 0 & \frac{\partial N_i}{\partial y} & 0 & \frac{\partial N_j}{\partial y} & 0 & \frac{\partial N_k}{\partial y} \\ \frac{\partial N_i}{\partial y} & \frac{\partial N_i}{\partial x} & \frac{\partial N_j}{\partial y} & \frac{\partial N_j}{\partial x} & \frac{\partial N_k}{\partial y} & \frac{\partial N_k}{\partial x} \end{bmatrix} \mathbf{U}^e \qquad (9.21)$$

$$= \mathbf{B}^e\mathbf{U}^e,$$

thus defining the matrix $\mathbf{B}^e$. The element stiffness matrix can be expressed in terms of $\mathbf{B}^e$ by

$$\mathbf{K}^e = \int\!\!\int_{R^e} (\mathbf{B}^e)^\mathsf{T}\mathbf{D}\mathbf{B}^e \; \mathrm{d}A \qquad (9.22)$$

To obtain the terms of the matrix we must revisit the ideas of chapter 8, sections 8.6.4 and 8.6.5 and general exercise 8.

The derivatives of the shape functions
The values of these terms can be obtained from the mapping of the general linear triangle from the standard triangle (figure 9.14). To recall some of the results of chapter 8, the shape functions on the master element and the mapping are

$$\hat{N}_1 = 1 - \xi - \eta \quad \hat{N}_2 = \xi \quad \hat{N}_3 = \eta,$$
$$T_e : x = \hat{N}_1 x_i + \hat{N}_2 x_j + \hat{N}_3 x_k$$
$$y = \hat{N}_1 y_i + \hat{N}_2 y_j + \hat{N}_3 y_k$$

The (linear) shape functions of the master element and the corresponding shape function of the general element have the same values at corresponding points,

$$N_i(x, y) = \hat{N}_1(\xi, \eta).$$

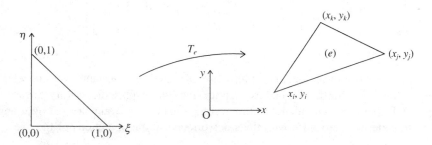

Figure 9.14 Mapping the master triangle onto the general linear triangle

The derivatives can then be obtained by

$$\frac{\partial N_i(x, y)}{\partial x} = \frac{\partial \hat{N}_1}{\partial \xi}\frac{\partial \xi}{\partial x} + \frac{\partial \hat{N}_1}{\partial \eta}\frac{\partial \eta}{\partial x}$$

$$= \frac{1}{|\mathbf{J}|}\left[\frac{\partial \hat{N}_1}{\partial \xi}\frac{\partial y}{\partial \eta} - \frac{\partial \hat{N}_1}{\partial \eta}\frac{\partial y}{\partial \xi}\right] \qquad \text{using equations (8.33)}$$

$$= \frac{1}{|\mathbf{J}|}[(-1)(y_k - y_i) - (-1)(-y_i + y_j)] \qquad \text{from the mapping}$$

$$= \frac{1}{|\mathbf{J}|}[(y_i - y_k) + (-y_i + y_j)]$$

$$= \frac{1}{|\mathbf{J}|}(y_j - y_k)$$

$$= \frac{b_i}{|\mathbf{J}|}$$

where the cyclic notation is used

$$\begin{array}{ll}
b_i = y_j - y_k & a_i = x_j - x_k \\
b_j = y_k - y_i & a_j = x_k - x_i \\
b_k = y_i - y_j & a_k = x_i - x_j.
\end{array} \qquad (9.23)$$

The Jacobian may be shown to be $|\mathbf{J}| = a_j b_k - a_k b_j$.

Exercise 9.6
Verify that

$$\frac{\partial N_j(x, y)}{\partial x} = \frac{b_j}{|\mathbf{J}|}, \qquad \frac{\partial N_k(x, y)}{\partial x} = \frac{b_k}{|\mathbf{J}|}$$

The other results are

$$\frac{\partial N_i(x, y)}{\partial y} = -\frac{a_i}{|\mathbf{J}|}, \qquad \frac{\partial N_j(x, y)}{\partial y} = -\frac{a_j}{|\mathbf{J}|}, \qquad \frac{\partial N_k(x, y)}{\partial y} = -\frac{a_k}{|\mathbf{J}|}.$$

Returning to the strains

$$\boldsymbol{\epsilon} = \mathbf{B}^e \mathbf{U}^e = \frac{1}{|\mathbf{J}|}\begin{bmatrix} b_i & 0 & b_j & 0 & b_k & 0 \\ 0 & -a_i & 0 & -a_j & 0 & -a_k \\ -a_i & b_i & -a_j & b_j & -a_k & b_k \end{bmatrix}\begin{bmatrix} U_{2i-1} \\ U_{2i} \\ U_{2j-1} \\ U_{2j} \\ U_{2k-1} \\ U_{2k} \end{bmatrix} \qquad (9.24)$$

It is now clear that the strains are obtained as constant over the element following from the original assumption that the displacements vary linearly.

To help gain confidence in this result, consider the strains for the right-angled triangle in figure 9.15, with the sides oriented in the directions of the axes.

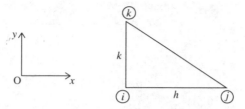

Figure 9.15   A right-angle triangle with sides in the directions of the axes

---

Exercise 9.7
Show that

$$b_i = -k \qquad a_i = h$$
$$b_j = k \qquad a_j = 0$$
$$b_k = 0 \qquad a_k = -h,$$

and that $|\mathbf{J}| = hk$.

Thus

$$\boldsymbol{\epsilon} = \mathbf{B}^e \mathbf{U}^e = \frac{1}{hk} \begin{bmatrix} -k & 0 & k & 0 & 0 & 0 \\ 0 & -h & 0 & 0 & 0 & h \\ -h & -k & 0 & k & h & 0 \end{bmatrix} \begin{bmatrix} U_{2i-1} \\ U_{2i} \\ U_{2j-1} \\ U_{2j} \\ U_{2k-1} \\ U_{2k} \end{bmatrix}$$

$$= \begin{bmatrix} (U_{2j-1} - U_{2i-1})/h \\ (U_{2k} - U_{2i})/k \\ (U_{2k-1} - U_{2i-1})/k + (U_{2j} - U_{2i})/h \end{bmatrix}$$

Hence the choice of linear element results in the strains being estimated by simple differences

$$\epsilon_{xx} = \frac{\partial u}{\partial x} = \left[\frac{U_{2j-1} - U_{2i-1}}{h}\right] = \left[\frac{u_j - u_i}{h}\right]$$

$$\epsilon_{yy} = \frac{\partial v}{\partial y} = \left[\frac{U_{2k} - U_{2i}}{k}\right] = \left[\frac{v_k - v_i}{k}\right] \qquad (9.25)$$

$$\gamma_{xy} = \frac{\partial u}{\partial y} + \frac{\partial v}{\partial x} = \left[\frac{U_{2k-1} - U_{2i-1}}{k}\right] + \left[\frac{U_{2j} - U_{2i}}{h}\right]$$

$$= \left[\frac{u_k - u_i}{k}\right] + \left[\frac{v_j - v_i}{h}\right]$$

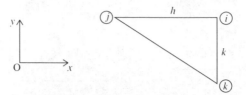

Figure 9.16   Another right-angled triangle with sides parallel to the axes

As a further example, and one which will help when working with the model problem later, consider the same triangle reflected about its hypotenuse (figure 9.16). In this case the matrix $\mathbf{B}^e$ is given by

$$\mathbf{B}^e = \frac{1}{hk} \begin{bmatrix} k & 0 & -k & 0 & 0 & 0 \\ 0 & h & 0 & 0 & 0 & -h \\ h & k & 0 & -k & -h & 0 \end{bmatrix}$$

As a result the strains for the element are

$$\epsilon_{xx} = \frac{\partial u}{\partial x} = \left[\frac{u_i - u_j}{h}\right]$$

$$\epsilon_{yy} = \frac{\partial v}{\partial y} = \left[\frac{v_i - v_k}{k}\right] \tag{9.26}$$

$$\gamma_{xy} = \frac{\partial u}{\partial y} + \frac{\partial v}{\partial x} = \left[\frac{u_i - u_k}{k}\right] + \left[\frac{v_i - v_j}{h}\right]$$

which is clearly the finite difference approximations for the derivatives when the triangle is oriented in this way.

Exercise 9.8
Verify $\mathbf{B}^e$ and the strain values given above.

Thus, once the displacements at the vertices of a triangular element are known, the strains and stresses may be calculated by

$$\boldsymbol{\epsilon} = \mathbf{B}^e\mathbf{U}^e \qquad \text{and} \qquad \boldsymbol{\sigma} = \mathbf{D}\boldsymbol{\epsilon}. \tag{9.27}$$

To conclude this section, the expression for the stiffness matrix and force vector will be obtained. Because the strains are constant, the integral for the element stiffness matrix may be evaluated as simply the integrand multiplied by the element area. From equation (9.20) and noting that $\mathbf{S}\Phi = \mathbf{B}$

$$\mathbf{K}^e = \int\int_{R^e} (\mathbf{B}^e)^{\mathrm{T}}\mathbf{D}\mathbf{B}^e \, \mathrm{d}A$$
$$= A(\mathbf{B}^e)^{\mathrm{T}}\mathbf{D}\mathbf{B}^e$$

where A is the area of the triangle.

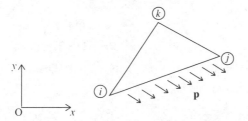

Figure 9.17   A load distributed on the side of a linear triangle

Consider forming the force vector for a uniformly distributed load **p** acting on the side of a triangle joining nodes i and j, S_{ij} say (figure 9.17). This is given as, see (9.20),

$$\mathbf{f}^e = \int_{S_{ij}} \boldsymbol{\phi}^{e\mathrm{T}} \mathbf{p} \; \mathrm{d}s.$$

Now the integration restricts the consideration to the side S_{ij}, where the basis functions are the element shape functions, and their integrals are just the areas of shaded triangles in figure 9.18.

$$
\begin{aligned}
\mathbf{f}^e &= \int_{S_{ij}} \begin{bmatrix} \phi_i & 0 & \phi_j & 0 & \phi_k & 0 \\ 0 & \phi_i & 0 & \phi_j & 0 & \phi_k \end{bmatrix}^{\mathrm{T}} \mathbf{p} \; \mathrm{d}s \\
&= \int_{S_{ij}} \begin{bmatrix} N_i & 0 & N_j & 0 & N_k & 0 \\ 0 & N_i & 0 & N_j & 0 & N_k \end{bmatrix}^{\mathrm{T}} \begin{bmatrix} p_x \\ p_y \end{bmatrix} \mathrm{d}s \\
&= \begin{bmatrix} \frac{l}{2} & 0 & \frac{l}{2} & 0 & 0 & 0 \\ 0 & \frac{l}{2} & 0 & \frac{l}{2} & 0 & 0 \end{bmatrix}^{\mathrm{T}} \begin{bmatrix} p_x \\ p_y \end{bmatrix} \\
&= \frac{l}{2} \begin{bmatrix} p_x \\ p_y \\ p_x \\ p_y \\ 0 \\ 0 \end{bmatrix}.
\end{aligned}
$$

where l is the length of the side S_{ij}. The distributed load is applied as a concentrated load, half on each of the two nodes i and j.

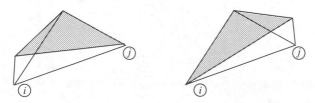

Figure 9.18   Shape functions on the side of a linear triangle

9.5.3 Summary

Element strain $\boldsymbol{\epsilon} = \mathbf{B}^e\mathbf{U}$ (9.28)

Element stress $\boldsymbol{\sigma} = \mathbf{DB}^e\mathbf{U}$ (9.29)

Element stiffness matrix $\mathbf{K}^e_c = A\mathbf{B}^{e\mathrm{T}}\mathbf{DB}^e$ (9.30)

For a uniform load on side S_{ij},

the element force vector is $\mathbf{f}^e_c = \dfrac{l}{2}\begin{bmatrix} p_x\ p_y\ p_x\ p_y\ 0\ 0 \end{bmatrix}^{\mathrm{T}}$ (9.31)

9.6 Working through the model problem with numerical values

We now return to the simple example shown in figure 9.12, which is to be used to illustrate the theory. Consider element 1; this fits the element of figure 9.15 with $h = k = 0.2$, so the element stiffness matrix is

$$\mathbf{K}^1_c = A\mathbf{B}^{1\mathrm{T}}\mathbf{DB}^1,$$

now $\quad \mathbf{DB}^1 = \dfrac{E}{0.2(1 - v^2)} \begin{bmatrix} 1 & v & 0 \\ v & 1 & 0 \\ 0 & 0 & \frac{1-v}{2} \end{bmatrix} \begin{bmatrix} -1 & 0 & 1 & 0 & 0 & 0 \\ 0 & -1 & 0 & 0 & 0 & 1 \\ -1 & -1 & 0 & 1 & 1 & 0 \end{bmatrix}$

$$= \dfrac{E}{0.2(1 - v^2)} \begin{bmatrix} -1 & -v & 1 & 0 & 0 & v \\ -v & -1 & v & 0 & 0 & 1 \\ -\frac{1-v}{2} & -\frac{1-v}{2} & 0 & \frac{1-v}{2} & \frac{1-v}{2} & 0 \end{bmatrix}$$

which eventually leads to

$$\mathbf{K}^1_c = \dfrac{E}{4(1 - v^2)} \begin{bmatrix} 3-v & 1+v & -2 & -1+v & -1+v & -2v \\ 1+v & 3-v & -2v & -1+v & -1+v & -2 \\ -2 & -2v & 2 & 0 & 0 & 2v \\ -1+v & -1+v & 0 & 1-v & 1-v & 0 \\ -1+v & -1+v & 0 & 1-v & 1-v & 0 \\ -2v & -2 & 2v & 0 & 0 & 2 \end{bmatrix}$$

If now the values for aluminium, $E = 7.10^{10}$, $v = 1/3$ are set in, the matrix becomes

$$\dfrac{21.10^{10}}{32} \begin{bmatrix} 8 & 4 & -6 & -2 & -2 & -2 \\ 4 & 8 & -2 & -2 & -2 & -6 \\ -6 & -2 & 6 & 0 & 0 & 2 \\ -2 & -2 & 0 & 2 & 2 & 0 \\ -2 & -2 & 0 & 2 & 2 & 0 \\ -2 & -6 & 2 & 0 & 0 & 6 \end{bmatrix}$$

The condensed stiffness matrix for element 1 refers to nodes 1, 2 and 3. So in expanded form referring to all the nodal variables, it is (omitting the constant factor for convenience)

$$
\begin{bmatrix}
8 & 4 & -6 & -2 & -2 & -2 & 0 & 0 & 0 & 0 & 0 & 0 \\
4 & 8 & -2 & -2 & -2 & -6 & 0 & 0 & 0 & 0 & 0 & 0 \\
-6 & -2 & 6 & 0 & 0 & 2 & 0 & 0 & 0 & 0 & 0 & 0 \\
-2 & -2 & 0 & 2 & 2 & 0 & 0 & 0 & 0 & 0 & 0 & 0 \\
-2 & -2 & 0 & 2 & 2 & 0 & 0 & 0 & 0 & 0 & 0 & 0 \\
-2 & -6 & 2 & 0 & 0 & 6 & 0 & 0 & 0 & 0 & 0 & 0 \\
0 & 0 & 0 & 0 & 0 & 0 & 0 & 0 & 0 & 0 & 0 & 0 \\
0 & 0 & 0 & 0 & 0 & 0 & 0 & 0 & 0 & 0 & 0 & 0 \\
0 & 0 & 0 & 0 & 0 & 0 & 0 & 0 & 0 & 0 & 0 & 0 \\
0 & 0 & 0 & 0 & 0 & 0 & 0 & 0 & 0 & 0 & 0 & 0 \\
0 & 0 & 0 & 0 & 0 & 0 & 0 & 0 & 0 & 0 & 0 & 0 \\
0 & 0 & 0 & 0 & 0 & 0 & 0 & 0 & 0 & 0 & 0 & 0
\end{bmatrix}
$$

The condensed stiffness matrix for element 3 is the same but, since it has nodes 3, 4 and 5, when expanded it fits into rows and columns 5, 6, 7, 8, 9 and 10 of the global matrix.

Element 2 is slightly different and fits the element of figure 9.16 with $h = k = 0.2$. It may be noted that $\mathbf{B}^2 = -\mathbf{B}^1$, so that the resulting stiffness matrix $\mathbf{K}_c^2$ is the same. However, the nodes used were 4, 3, 2 corresponding to i, j, k, so that if the order is changed to 2, 3, 4, which is more convenient for setting into the global stiffness matrix by hand, $\mathbf{K}^2$ is, without the constant factor,

$$
\begin{bmatrix}
0 & 0 & 0 & 0 & 0 & 0 & 0 & 0 & 0 & 0 & 0 & 0 \\
0 & 0 & 0 & 0 & 0 & 0 & 0 & 0 & 0 & 0 & 0 & 0 \\
0 & 0 & 2 & 0 & 0 & 2 & -2 & -2 & 0 & 0 & 0 & 0 \\
0 & 0 & 0 & 6 & 2 & 0 & -2 & -6 & 0 & 0 & 0 & 0 \\
0 & 0 & 0 & 2 & 6 & 0 & -6 & -2 & 0 & 0 & 0 & 0 \\
0 & 0 & 2 & 0 & 0 & 2 & -2 & -2 & 0 & 0 & 0 & 0 \\
0 & 0 & -2 & -2 & -6 & -2 & 8 & 4 & 0 & 0 & 0 & 0 \\
0 & 0 & -2 & -6 & -2 & -2 & 4 & 8 & 0 & 0 & 0 & 0 \\
0 & 0 & 0 & 0 & 0 & 0 & 0 & 0 & 0 & 0 & 0 & 0 \\
0 & 0 & 0 & 0 & 0 & 0 & 0 & 0 & 0 & 0 & 0 & 0 \\
0 & 0 & 0 & 0 & 0 & 0 & 0 & 0 & 0 & 0 & 0 & 0 \\
0 & 0 & 0 & 0 & 0 & 0 & 0 & 0 & 0 & 0 & 0 & 0 \\
0 & 0 & 0 & 0 & 0 & 0 & 0 & 0 & 0 & 0 & 0 & 0 \\
0 & 0 & 0 & 0 & 0 & 0 & 0 & 0 & 0 & 0 & 0 & 0
\end{bmatrix}
$$

The stiffness matrix for element 4 is similar, but refers to nodes 4, 5 and 6, and so has the same pattern of numbers, but moved down the diagonal of the 12×12 structure of the global stiffness matrix.

Exercise 9.9
Form the stiffness matrix for element 4.

The four stiffness matrices may be added to form the global matrix which is $21.10^{10}/32$ multiplying

$$
\begin{bmatrix}
8 & 4 & -6 & -2 & -2 & -2 & 0 & 0 & 0 & 0 & 0 & 0 \\
4 & 8 & -2 & -2 & -2 & -6 & 0 & 0 & 0 & 0 & 0 & 0 \\
-6 & -2 & 8 & 0 & 0 & 4 & -2 & -2 & 0 & 0 & 0 & 0 \\
-2 & -2 & 0 & 8 & 4 & 0 & -2 & -6 & 0 & 0 & 0 & 0 \\
-2 & -2 & 0 & 4 & 16 & 4 & -12 & -4 & -2 & -2 & 0 & 0 \\
-2 & -6 & 4 & 0 & 4 & 16 & -4 & -4 & -2 & -6 & 0 & 0 \\
0 & 0 & -2 & -2 & -12 & -4 & 16 & 4 & 0 & 4 & -2 & -2 \\
0 & 0 & -2 & -6 & -4 & -4 & 4 & 16 & 4 & 0 & -2 & -6 \\
0 & 0 & 0 & 0 & -2 & -2 & 0 & 4 & 8 & 0 & -6 & -2 \\
0 & 0 & 0 & 0 & -2 & -6 & 4 & 0 & 0 & 8 & -2 & -2 \\
0 & 0 & 0 & 0 & 0 & 0 & -2 & -2 & -6 & -2 & 8 & 4 \\
0 & 0 & 0 & 0 & 0 & 0 & -2 & -6 & -2 & -2 & 4 & 8
\end{bmatrix}.
$$

The force vector comes from the distributed load on the side of element 4 joining nodes 5 and 6. This is in the downward direction and is equivalent to a load $10^6 \times 0.2 \times \frac{1}{2}$ N. Thus the global force vector is

$$
10^5 \begin{bmatrix} 0 & 0 & 0 & 0 & 0 & 0 & 0 & 0 & 0 & -1 & 0 & -1 \end{bmatrix}^{\mathrm{T}}
$$

The boundary conditions are that because of symmetry there is restraint in the x-direction at nodes $1, 3$ and 5, and the support at node 2 means that the y-displacement is zero. Thus the variables numbered $1, 5, 9$ and 4 are all zero: so the reduced set of equations is obtained by leaving out these rows and columns. The stiffness matrix and force vector are then given by

$$
\begin{bmatrix}
8 & -2 & -6 & 0 & 0 & 0 & 0 & 0 \\
-2 & 8 & 4 & -2 & -2 & 0 & 0 & 0 \\
-6 & 4 & 16 & -4 & -4 & -6 & 0 & 0 \\
0 & -2 & -4 & 16 & 4 & 4 & -2 & -2 \\
0 & -2 & -4 & 4 & 16 & 0 & -2 & -6 \\
0 & 0 & -6 & 4 & 0 & 8 & -2 & -2 \\
0 & 0 & 0 & -2 & -2 & -2 & 8 & 4 \\
0 & 0 & 0 & -2 & -6 & -2 & 4 & 8
\end{bmatrix}
\begin{bmatrix}
v_1 \\ u_2 \\ v_3 \\ u_4 \\ v_4 \\ v_5 \\ u_6 \\ v_6
\end{bmatrix}
= \frac{32.10^{-5}}{21}
\begin{bmatrix}
0 \\ 0 \\ 0 \\ 0 \\ 0 \\ -1 \\ 0 \\ -1
\end{bmatrix},
$$

which has the solution

$$
10^{-5}[-0.392 \quad 0.103 \quad -0.558 \quad 0.101 \quad -0.411 \quad -0.819 \quad 0.076 \quad -0.716]
$$

this is illustrated in figure 9.19, with the displacements at each node shown. The values shown are to be multiplied by 10^{-5}.

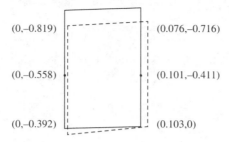

(0,–0.819)   (0.076,–0.716)

(0,–0.558)   (0.101,–0.411)

(0,–0.392)   (0.103,0)

Figure 9.19   The solution showing the displaced shape dotted

Now that the displacements have been calculated, the strains and stresses for each element may be derived. For element 1, using (9.25)

$$\epsilon_{xx} = \qquad [(0.103 - 0.000)\,10^{-5}]/0.2 = 0.515 \times 10^{-5}$$
$$\epsilon_{yy} = [(-0.558 - (-0.392))\,10^{-5}]/0.2 = -0.83 \times 10^{-5} \qquad (9.32)$$
$$\gamma_{xy} = \qquad [0 + (0.392)10^{-5}]/0.2 = 1.96 \times 10^{-5}$$

The stresses follow, from (9.6)

$$
\begin{bmatrix} \sigma_{xx} \\ \sigma_{yy} \\ \sigma_{xy} \end{bmatrix}
= \frac{E}{1 - \nu^2}
\begin{bmatrix} 1 & \nu & 0 \\ \nu & 1 & 0 \\ 0 & 0 & \frac{1-\nu}{2} \end{bmatrix}
\begin{bmatrix} \epsilon_{xx} \\ \epsilon_{yy} \\ \gamma_{xy} \end{bmatrix}
$$

$$
= \frac{21}{8} 10^5
\begin{bmatrix} 3 & 1 & 0 \\ 1 & 3 & 0 \\ 0 & 0 & 1 \end{bmatrix}
\begin{bmatrix} 0.515 \\ -0.83 \\ 1.96 \end{bmatrix}
$$

$$
=
\begin{bmatrix} 1.88 \times 10^5 \\ -5.18 \times 10^5 \\ 5.15 \times 10^5 \end{bmatrix}
$$

Exercise 9.10
Calculate the strains and stresses for element 2.

Appendix A

To show that $\mathrm{tr}[\mathbf{S}^{\mathrm{T}}(\boldsymbol{\sigma}\mathbf{w}^{\mathrm{T}})] = \mathbf{w}^{\mathrm{T}}\mathbf{S}^{\mathrm{T}}\boldsymbol{\sigma} + \boldsymbol{\sigma}^{\mathrm{T}}\mathbf{S}\mathbf{w}$.

Expanding the left-hand side,

$$\text{tr}\left[\mathbf{S}^{\mathsf{T}}(\boldsymbol{\sigma}\mathbf{w}^{\mathsf{T}})\right] = \text{tr}\left\{ \begin{bmatrix} \frac{\partial}{\partial x} & 0 \\ 0 & \frac{\partial}{\partial y} \\ \frac{\partial}{\partial y} & \frac{\partial}{\partial x} \end{bmatrix}^{\mathsf{T}} \begin{bmatrix} \sigma_{xx}w_1 & \sigma_{xx}w_2 \\ \sigma_{yy}w_1 & \sigma_{yy}w_2 \\ \sigma_{xy}w_1 & \sigma_{xy}w_2 \end{bmatrix} \right\}$$

$$= \frac{\partial}{\partial x}(\sigma_{xx}w_1 + \sigma_{xy}w_2) + \frac{\partial}{\partial y}(\sigma_{xy}w_1 + \sigma_{yy}w_2) \qquad (9.33)$$

Differentiating

$$= w_1\left(\frac{\partial \sigma_{xx}}{\partial x} + \frac{\partial \sigma_{xy}}{\partial y}\right) + w_2\left(\frac{\partial \sigma_{yy}}{\partial y} + \frac{\partial \sigma_{xy}}{\partial x}\right)$$

$$+ \sigma_{xx}\frac{\partial w_1}{\partial x} + \sigma_{yy}\frac{\partial w_2}{\partial y} + \sigma_{xy}\left(\frac{\partial w_1}{\partial y} + \frac{\partial w_2}{\partial x}\right).$$

By direct expansion the two terms on the right-hand side are:

$$\mathbf{w}^{\mathsf{T}}\mathbf{S}^{\mathsf{T}}\boldsymbol{\sigma} = w_1\left(\frac{\partial \sigma_{xx}}{\partial x} + \frac{\partial \sigma_{xy}}{\partial y}\right) + w_2\left(\frac{\partial \sigma_{yy}}{\partial y} + \frac{\partial \sigma_{xy}}{\partial x}\right)$$

$$\boldsymbol{\sigma}^{\mathsf{T}}\mathbf{S}\mathbf{w} = \sigma_{xx}\frac{\partial w_1}{\partial x} + \sigma_{yy}\frac{\partial w_2}{\partial y} + \sigma_{xy}\left(\frac{\partial w_1}{\partial y} + \frac{\partial w_2}{\partial x}\right).$$

Thus the result is shown.

Appendix B

To show that

$$\int\int_R \text{tr}\left[\mathbf{S}^{\mathsf{T}}(\boldsymbol{\sigma}\mathbf{w}^{\mathsf{T}})\right]\mathrm{d}A = \int_S \mathbf{w}^{\mathsf{T}}\mathbf{M}\boldsymbol{\sigma}\ \mathrm{d}s.$$

Consider the integral over R as an iterated integral, and the first integration taking place over x. The surface S is divided into two parts $x = g_1(y)$ and $x = g_2(y)$ by lines $y = a$ and $y = b$ parallel to the x-axis. This, and the relation between a small segment of the boundary δs and its projection in the y-direction, δy, are shown in figure 9.20.

Given a general function $f(x, y)$ we have

$$\int\int_R \frac{\partial}{\partial x} f(x, y)\ \mathrm{d}A = \int_{y=a}^{y=b}\left[\int_{x=g_1(y)}^{x=g_2(y)} \frac{\partial}{\partial x} f(x, y)\ \mathrm{d}x\right]\mathrm{d}y$$

$$= \int_{y=a}^{y=b} [f(x, y)]_{x=g_1(y)}^{x=g_2(y)}\ \mathrm{d}y$$

$$= \int_a^b [f(g_2(y), y) - f(g_1(y), y)]\ \mathrm{d}y$$

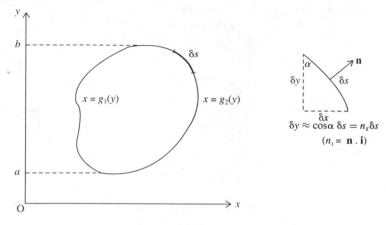

Figure 9.20   Projecting the boundary

$$= \int_a^b f(g_2(y), y)n_x \; ds - \int_a^b f(g_1(y), y)n_x \; ds$$

$$= \int_S f(x, y)n_x \; ds$$

allowing for the outward direction of the normal. Similarly,

$$-\int\int_R \frac{\partial}{\partial y} f(x, y) \; dA = \int_S f(x, y)n_y \; ds.$$

Thus in transforming from a domain integral to a boundary integral, $\frac{\partial}{\partial x}$ and $\frac{\partial}{\partial y}$ are replaced by n_x and n_y.

From the expansion (9.33)

$$\int\int_R \text{tr}\big[S^T(\sigma w^T)\big] \; dA = \int\int_R \left[\frac{\partial}{\partial x}(\sigma_{xx}w_1 + \sigma_{xy}w_2) + \frac{\partial}{\partial x}(\sigma_{xx}w_1 + \sigma_{xy}w_2)\right] dA$$

$$= \int_S \big[n_x(\sigma_{xx}w_1 + \sigma_{xy}w_2) + n_y(\sigma_{xx}w_1 + \sigma_{xy}w_2)\big] \; ds$$

which may be seen to be

$$\int_S \text{tr}\big[M^T(\sigma w^T)\big] \; ds = \int_S \big[w_1(\sigma_{xx}n_x + \sigma_{xy}n_y) + w_2(\sigma_{yy}n_y + \sigma_{xy}n_x)\big] \; ds$$

$$= \int_S w^T M^T \sigma \; ds$$

General exercises for chapter 9

1. Consider the same problem discussed in section 9.4 but discretised using the mesh in figure 9.21.

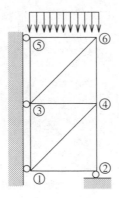

Figure 9.21   Exercise 1

Show that the coefficients of the corresponding stiffness matrix are different from those given in section 9.6 and therefore that a different solution can be expected. Check that the displacements given in section 9.6 do not satisfy the new system of equations. Comment.

2. Obtain the stiffness matrix of an equilateral triangle (figure 9.22) of size $l = 1$ m, $E = 7 \times 10^{10}$ and $\nu = 1/3$.

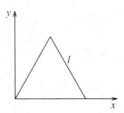

Figure 9.22   Exercise 2

3. Using the elemental stiffness matrix obtained in exercise 2, assemble the global stiffness of the mesh shown in figure 9.23 and obtain the displacements of the only free node and the stresses in each element.

4. Consider the body force term in equation (9.20). Show that if $\mathbf{b}$ is constant as in the case of self-weight forces where $\mathbf{b} = -\rho g \mathbf{j}$, then the resulting force vector for a general linear triangular element of area A and thickness t is:

$$\mathbf{f}^e = -\frac{1}{3} A t \rho g \begin{bmatrix} 0 \\ 1 \\ 0 \\ 1 \\ 0 \\ 1 \end{bmatrix}$$

Obtain the self-weight force vector for exercise 1, taking $\rho = 5 \times 10^3$.

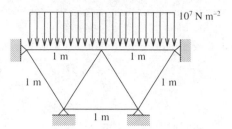

Figure 9.23   Exercise 3

5. Consider a linearly varying distributed load per unit length $\mathbf{p} = \mathbf{p}_i + \frac{x}{l}(\mathbf{p}_j - \mathbf{p}_i)$ over an element side of length l joining nodes i to j as shown in figure 9.24. (Note that $\mathbf{p}_i$ need not be parallel to $\mathbf{p}_j$.)

Show that the resulting force vector is

$$\mathbf{f}^e = \int_0^l \mathbf{\Phi}^{e\mathrm{T}} \mathbf{p}\, \mathrm{d}x = \frac{lt}{6} \begin{bmatrix} 2\mathbf{p}_i + \mathbf{p}_j \\ 2\mathbf{p}_j + \mathbf{p}_i \\ 0 \\ 0 \end{bmatrix}$$

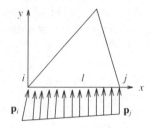

Figure 9.24   Exercise 5

6. Consider a six-noded quadratic triangular element with uniform load $\mathbf{p} = [p_x, p_y]^\mathrm{T}$ acting on the side joining nodes 1, 2 and 3 of length l (figure 9.25).

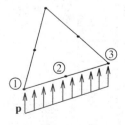

Figure 9.25   Exercise 6

(i) Write down the shape function matrix $\mathbf{\Phi}^e$ corresponding to the line integral for the load.

(ii) Show that the element force vector is given by:

$$\mathbf{f}^e = \frac{l}{6}[p_x, p_y, 4p_x, 4p_y, p_x, p_y, 0, 0, \ldots, 0]^\mathsf{T}$$

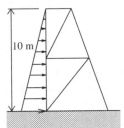
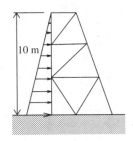
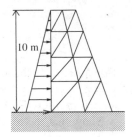

Figure 9.26   Exercise 7

7. Using the equation derived in exercise 5, obtain the global force vector due to the water pressure behind a dam of height 10 m and thickness $t = 1$ m. Use 2, 3 and 4 elements along the height of the dam as shown in figure 9.26. The pressure at the bottom of the dam is $12 \times 10^6 \, \mathrm{N\,m^2}$.

8. Generally, a finite element analysis of a given problem provides only an approximation to the exact analytical solution. In exceptional cases, however, the exact stresses are uniform and the solution obtained using linear triangles becomes exact. Consider, for instance, the block resting on smooth frictionless surfaces shown in figure 9.27 and a simple discretisation using two triangles. Assemble the global stiffness matrix of the problem and the force vector and obtain nodal displacements and stresses in all elements. Show that these stresses are uniform and equal to the exact values.

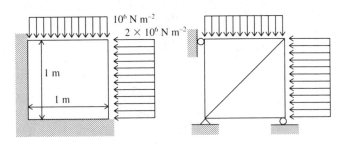

Figure 9.27   Exercise 8

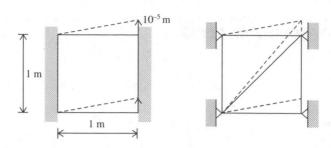

Figure 9.28   Exercise 9

9. Consider a simple shear test where a square block of material is confined between two rigid plates, which are kept at a constant distance, but displaced relative to each other. In order to simulate this test with finite elements, consider the meshes shown in figure 9.28 where the nodes on the left are fixed and the nodes on the right-hand side are fixed in the x-direction and subject to a prescribed vertical displacement of 10^{-5} m.

Using $E = 7 \times 10^{10}$ and $\nu = 1/3$ obtain the stresses in each element. (Note that since all the nodal displacements are known, there is no need to evaluate the stiffness matrix or solve any equations.) Discuss the need to use a larger number of elements in the mesh.

10. In an effort to analyse a pure bending problem for a material with $E = 7 \times 10^{10}$ and $\nu = 0$, the mesh in figure 9.29 is used. The nodes on the left are fixed, while the right-hand side section is rotated an angle $\theta = 10^{-5}$ by fixing the nodes in the y-direction and prescribing a linearly varying displacement in the x-direction as $u = \theta y$.

Obtain the resulting stresses in the right-hand side section and plot these against the exact values $\sigma_{xx} = E\theta y/l$. Solve the same exercise using three and four pairs of elements in the vertical direction. Discuss the need for using a large number of elements.

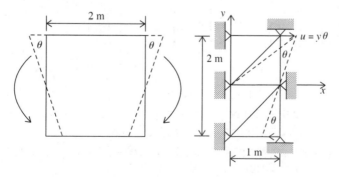

Figure 9.29   Exercise 10

Answers to exercises

Chapter 1

1.2

$$\frac{AE}{l}\begin{bmatrix} 1 & 0 & -1 & 0 \\ 0 & 0 & 0 & 0 \\ -1 & 0 & 1 & 0 \\ 0 & 0 & 0 & 0 \end{bmatrix}\begin{bmatrix} u_2 \\ v_2 \\ u_4 \\ v_4 \end{bmatrix} = \begin{bmatrix} U_2^4 \\ V_2^4 \\ U_4^4 \\ V_4^4 \end{bmatrix}$$

1.3 $u_2' = v_2$, $v_2' = -u_2$, $U_2' = V_2'$, $V_2' = -U_2'$

1.4 $U_3'^5 + U_4'^5 = 0$, $U_4'^5 = EA(u_4' - u_3)/l$, $V_3'^5 = V_4'^5 = 0$.

1.5 Verifying $\mathbf{TT}^{\mathrm{T}} = \mathbf{T}^{\mathrm{T}}\mathbf{T} = \mathbf{I}$.

General exercises

2. Element at 60°, nodes 1,2:

$$\frac{AE}{4l}\begin{bmatrix} 1 & \sqrt{3} & -1 & \sqrt{3} \\ \sqrt{3} & 3 & -\sqrt{3} & -3 \\ -1 & -\sqrt{3} & 1 & \sqrt{3} \\ \sqrt{3} & -3 & \sqrt{3} & 3 \end{bmatrix}\begin{bmatrix} u_1 \\ v_1 \\ u_2 \\ v_2 \end{bmatrix} = \begin{bmatrix} U_1 \\ V_1 \\ U_2 \\ V_2 \end{bmatrix}$$

3. $u_2 = 2 + 2\sqrt{2}$, $v_2 = 1$, $u_3 = 1$, $u_4 = 3 + 2\sqrt{2}$, $v_4 = 0$ in mm. Forces 1000 N or $-1000\sqrt{2}$ N.

4. $u_3 = -1$, $v_3 = -2\sqrt{2} - 1 = v_4$, $u_4 = 1$. Forces 1000 N or $-1000\sqrt{2}$ N.

5. Element 7:

$$\frac{AE}{2\sqrt{2}l}\begin{bmatrix} 1 & -1 & -1 & 1 \\ -1 & 1 & 1 & -1 \\ -1 & 1 & 1 & -1 \\ 1 & -1 & -1 & 1 \end{bmatrix}\begin{bmatrix} u_4 \\ v_4 \\ u_5 \\ v_5 \end{bmatrix} = \begin{bmatrix} U_4^7 \\ V_4^7 \\ U_5^7 \\ V_5^7 \end{bmatrix}$$

6. $u_2 = 0$, $u_3 = 2\sqrt{2} + 1$, $v_3 = -1$ mm.

7. $2\sqrt{2}$ N, -2.0 N, 2.0 N.

8. (a) $u_2 = 0$, $v_2 = -0.625$ mm. (b) $1.2\sqrt{2}$ mm.

9. A : $X = U_1^1 + U_1^3$, $Y = V_1^1 + V_1^3$. B: $0 = U_2^1 + U_2^2$, $Y = V_2' + V_2^2$. C: $0 = U_3^2 + U_3^3$, $-100 = V_3^2 + V_3^3$

Chapter 2

2.6 $u(1/3) = 7/18$, $u(2/3) = 13/18$, both exact.

2.7 -0.0185

2.8 $u_1 = 9/16$

General exercises

1. (i) 1.1752, (ii) 1.1379, (ii).

2. At $x = l/4, 3l/4$ FE gives $-.0065wl^4/EI$ and exact solution is $-.0093wl^4/EI$. At $x = l/2$ both are $-5wl^4/384EI$.

Chapter 3

3.5 $u_2 = 11/36$, $u_3 = 20/36$, $u_4 = 27/36$.

3.6 $u_2 = 7/18$, $u_3 = 14/18$, exact.

3.7 $V(u) = T_B u_B + \int_0^1 u'^2 - (x+u)\,dx$, $u_2 = 5/36$, $u_3 = 8/36$, $9/36$.

General exercises

1. $u_2 = 19/64$, $u_3 = 36/64$, $u_4 = 51/64$.

2. $u_2 = 11/36$, $u_3 = 20/36$, $u_4 = 27/36$.

3. $u_2 = 19/64$, $u_3 = 36/64$.

4. $u_2 = -5/256$, $u_3 = -8/256$, $u_4 = -7/256$.

6. $u_2 = 0.8384$, $u_3 = 0.7298$, $u_4 = 0.6672$, $u_5 = 0.6468$.

7. $u_2 = -0.0154$, $u_3 = -0.1457$, $u_4 = -0.405$.

8. $u_1 = 12$, $u_2 = 20$, $u_3 = 24$.

Chapter 4

4.2 $$\begin{bmatrix} 0 & 0 & 0 & 0 \\ 0 & 0 & 0 & 0 \\ 0 & 0 & 2/h & -2/h \\ 0 & 0 & -2/h & 2/h \end{bmatrix}, \begin{bmatrix} 0 \\ 0 \\ h/2 \\ h/2 \end{bmatrix}$$

4.6
$$
\begin{bmatrix}
12 & -12 & & & & \\
-12 & 24 & -12 & & & \\
& -12 & 24 & -12 & & \\
& & -12 & 20 & -8 & \\
& & & -8 & -16 & -8 \\
& & & & -8 & 8
\end{bmatrix},
\begin{bmatrix}
1/12 - T_B \\
1/6 \\
1/6 \\
5/24 \\
1/4 \\
1/8 + T_C
\end{bmatrix}
$$

4.7

(a) (i) 3, (ii) 3, (iii) 3.

(b) (i) $(E + 1)(E + 1)$, (ii) & (iii) $3(E + 1) - 2$

4.8 $u_1 = -27/36$, $u_2 = -16/36$, $u_3 = -7/36$, $u_4 = 0$.

General exercises

1.
$$
\frac{2}{h}
\begin{bmatrix}
2 & -1 & & & \\
-1 & 2 & -1 & & \\
& -1 & 2 & -1 & \\
& & \ddots & & -1 \\
& & & -1 & 2
\end{bmatrix},
-h^2
\begin{bmatrix}
1 \\
2 \\
3 \\
\vdots \\
N-2
\end{bmatrix}
$$

2.
$$
\begin{bmatrix}
\frac{4}{h} + \frac{4h}{3} & -\frac{2}{h} + \frac{h}{3} & & \\
-\frac{2}{h} + \frac{h}{3} & \frac{4}{h} + \frac{4h}{3} & & \\
& & \ddots & -\frac{2}{h} + \frac{h}{3} \\
& & -\frac{2}{h} + \frac{h}{3} & \frac{4}{h} + \frac{4h}{3}
\end{bmatrix},
\begin{bmatrix}
\frac{2}{h} - \frac{h}{3} \\
0 \\
\vdots \\
0
\end{bmatrix}
$$

3. $u_2 = 90$, $u_3 = 80$, $u_4 = 70$, $u_5 = 60$, $q = 40$.

4. $u_2 = -0.0018$, $u_3 = -0.0068$, $u_4 = -0.1993$, $u_5 = -0.4075$.

5. Both methods give $u_2 = 70$, $u_3 = 50$, $u_4 = 46$.

Chapter 5

5.5 $\nabla u \approx (u_j - u_i)\hat{\mathbf{i}}/l + (u_k - u_i)\hat{\mathbf{j}}/l$. Intersection of perpendicular bisection of the sides.

5.9 $K(7, 2) = K(7, 2) + K_{ij}$, $K(7, 11) = K(7, 11) + K_{ik}$, etc.

General exercises

1. Set $q^* = 0$ in (5.2).
2. $u_1 = 9.79$, $u_2 = 9.18$, $u_3 = 6.91$, $u_4 = 4.23$.
3. $N_i = (\sqrt{3}l - \eta - \sqrt{3}\xi)/(2l\sqrt{3})$, $N_i = (\sqrt{3}l - \eta + \sqrt{3}\xi)/(2l\sqrt{3})$, $N_k = \eta/(l\sqrt{3})$.
4.

$$\frac{K}{2\sqrt{3}}\begin{bmatrix} 2 & -1 & -1 \\ -1 & 2 & -1 \\ -1 & -1 & 2 \end{bmatrix}$$

5. 37.16, 67.56, 68.24.
6. 350/3, 250/3.
7. Heat lost through convection 20.58 kW. Output at nodes with prescribed temperature 9.43 kW. Conservation to within rounding accuracy.
8. 81.58, 63.08, 2 × 923.08 W.
9. $u_1 = 50.2$, $u_2 = 49.8$, $u_3 = 80.8$, $u_4 = 79.2$, $u_5 = 91.1$, $u_6 = 105.1$, $u_7 = 94.3$, $u_8 = 97.7$.

Chapter 6

6.1 $-3 + 3x - x^2$, 0
6.2 $r(1/2) = 0$ implies $\alpha = -5/7$.
6.3 (b) $\int_0^1 (uu'' + 3u')v\,dx = 0 \ \forall v: \ v(0) = v(1) = 0$.

6.5 $\dfrac{\partial v}{\partial x}\dfrac{\partial u}{\partial x} + \dfrac{\partial v}{\partial y}\dfrac{\partial u}{\partial y}$

6.7 $u(0) = 0$ and $\int_0^1 (u'v' + auv)\,dx = \int_0^1 fv\,dx + (bu(1) + c)v(1) \ \forall v: \ v(0) = 0$.
6.12 Consider B_2 for $u = 1$, $u = e^{2x}$, opposite sign.
6.14 Two forms differ by a constant.

General exercises

1. (a) $\cos x + \dfrac{1 - \cos 1}{\sin 1}\sin x - 1$, (b) $-2\alpha + \alpha x(1 - x) + 1$, (c) $\alpha = 4/7$,
   (d) $\alpha = 55/101\cdot$
2. (a) $\dfrac{\sinh x}{\cosh 1} + x$. (c) $u = 7x/4$, $u = 551x/347 + 60x^2/347$.
3. $B(u, v) = \int_0^1 (-u'v' + au'v + buv)\,dx$, $L(v) = \int_0^1 -fv\,dx$.

4. $w(0) = w(l) = 0$ and $\int_0^l EIw''v'' \, dx = \int_0^l fv \, dx \ \forall v : \ v(0) = v(l) = 0$.
     $V(w) = \frac{1}{2}\int_0^l EI(w'')^2 \, dx - \int_0^l fw \, dx$. Since $w''(x)$ must be bounded, piecewise
     linear trial functions lack sufficient smoothness.

5.   It is possible but awkward.

6. $u : u(0) = u^*$ and $\int_A^B ku'v' \, dx + hu(B)v(B) = u_\infty v(B) \ \forall v : \ v(0) = 0$, where
     $k = k_1, k_2, k_3$ appropriately.

7. $B(u, w) = \iint_R (\nabla u \cdot \nabla w - k^2 uw) \, dA$, $L(w) = \int_{S_2} v^* w \, ds$. B is not positive
     definite.

8. $u = u_1^*$ on S_1 and u_2^* on S_2, and $\iint_R k \nabla u \cdot \nabla v \, dA = 0 \ \forall v : \ v = 0$ on $S_1 \cup S_2$,
     where $k = k_1, k_2$ appropriately.

Chapter 7

7.1   (a) Yes, (b) Yes, (c) No, (d) Yes.

7.2 $N_j(x) = 4(x - x_i)(x_k - x)/h^2$, $N_k(x) = 2(x - x_i)(x - x_j)/h^2$.

7.4 N_i, N_j, N_k span quadratics and hence $u(x) = 1$.

7.13 $\alpha_1 = 5/18$, error: 0, -0.0081, 0.0003, 0.0080, 0.

7.14 $u : u(0) = 0$ and $\int_0^1 (u'v' - uv) \, dx = \int_0^1 xv \, dx$, $\forall v : \ v(0) = 0$, $u = 0.5x$,
      $u = 0.9856x - 0.4317x^2$.

7.15 Show that $K_{rc} = K_{cr}$.

7.17

$$\frac{1}{h}\begin{bmatrix} 1 & -1 \\ -1 & 1 \end{bmatrix}, \quad \frac{1}{6h}\begin{bmatrix} 7 & -8 & 1 \\ -8 & 16 & -8 \\ 1 & -8 & 7 \end{bmatrix}$$

General exercises

1. $u : u(0) = 0$ and $\int_0^1 (u'v' - uv) \, dx - u(1)v(1) = -\int_0^1 xv \, dx + v(1) \ \forall v :$
     $v(0) = 1$. $u = -\frac{199}{77}x + \frac{60}{77}x^2$.

2.   (b) $N_i(x) = (x - x_j)^2[h + 2(x - x_i)]/h^3$, $M_i(x) = (x - x_j)^2(x - x_i)/h^2$.

     (c)
$$\mathbf{K}_c^e = \begin{bmatrix} \int_{R^e} N_i'' N_i'' \, dx & \int_{R^e} M_i'' N_i'' \, dx & \cdots \\ \int_{R^e} N_i'' M_i'' \, dx & \int_{R^e} M_i'' M_i'' \, dx & \cdots \\ \vdots & \vdots & \ddots \end{bmatrix}$$

4.   (b) Result based on $\sum_i N_i = 1$ and holds generally.

5.   (b) Will not hold for quadratic elements.

6. Row sums, $-bh/2$, column sums, $(\pm a/h) - (bh/2)$. Similar results hold for quadratics.

7. $\phi = 5.0, 3.667, 3.667, 2.833$ and see section 7.8.

Chapter 8

8.2 $f_j^e = \int_{-1}^{1}(x_j + h\xi)(1 - \xi^2)h \ d\xi$.

8.4 6, exact.

8.5 4/9, exact.

8.6 5/24, exact.

8.15 $x = 2\hat{N}_2 + \hat{N}_3$, $y = \hat{N}_3$.

8.19 Yes.

8.20 $\dfrac{\partial \eta}{\partial x} = -\dfrac{1}{|\mathbf{J}|}\dfrac{\partial}{\partial \xi}\left[\sum_{1}^{m}\hat{N}_i(\xi, \eta)y_i\right]$

General exercises

1. $e(1) = 0$ implies $w_1 + w_2 = \int_{-1}^{1} 1 \ dx = 2$; similarly, $w_1x_1 + w_2x_2 = 0$, $w_1x_1^2 + w_2x_2^2 = 2/3$, $w_1x_1^3 + w_2x_2^3 = 0$.

4. Only if P_2, P_4 and P_6 are mid-points of the sides.

8.
$$\mathbf{K} = \frac{1}{|\mathbf{J}|}\begin{bmatrix} b_ib_i + a_ia_i & b_ib_j + a_ia_j & b_ib_k + a_ia_k \\ b_jb_i + a_ja_i & b_jb_j + a_ja_j & b_jb_k + a_ja_k \\ b_kb_i + a_ka_i & b_kb_j + a_ka_j & b_kb_k + a_ka_k \end{bmatrix}$$

9. Continuity for both the geometry and the problem variable.

10. No.

Chapter 9

9.1
$$\frac{E}{1 - \nu^2}\begin{bmatrix} \frac{\partial^2}{\partial x^2} + \frac{1-\nu}{2}\frac{\partial^2}{\partial y^2} & \frac{1+\nu}{2}\frac{\partial^2}{\partial x \partial y} \\ \frac{1+\nu}{2}\frac{\partial^2}{\partial x \partial y} & \frac{\partial^2}{\partial y^2} + \frac{1-\nu}{2}\frac{\partial^2}{\partial x^2} \end{bmatrix}$$

9.10 $\boldsymbol{\epsilon} = [0.505, -2.055, 0.745]^{\mathrm{T}}10^{-5}$. $\boldsymbol{\sigma} = [1.42, -14.86, 1.96]^{\mathrm{T}}10^5$.

General exercises

2.

$$\mathbf{K} = \frac{7\sqrt{3}}{32}10^{10}\begin{bmatrix} 10 & 2\sqrt{3} & -8 & 0 & -2 & -2\sqrt{3} \\ 2\sqrt{3} & 6 & 0 & 0 & -2\sqrt{3} & -6 \\ -8 & 0 & 10 & -2\sqrt{3} & -2 & 2\sqrt{3} \\ 0 & 0 & -2\sqrt{3} & 6 & 2\sqrt{3} & -6 \\ -2 & -2\sqrt{3} & -2 & 2\sqrt{3} & 4 & 0 \\ -2\sqrt{3} & -6 & 2\sqrt{3} & -6 & 0 & 12 \end{bmatrix}$$

3. $u_4 = 0$, $v_4 = -0.11$ mm, $\boldsymbol{\sigma}^1 = (7\sqrt{3}/8)10^7[-0.22, -0.66, 0]^T$,
     $\boldsymbol{\sigma}^2 = (7\sqrt{3}/8)10^7[0.11, 0.33, -0.11\sqrt{3}]^T$.

7. $\frac{5}{6}[30, 0, 36, 0, 6, 0, ...]^T$, $(10/18)[32, 0, 48, 0, 24, 0, 4, 0, ...]^T$,
     $(5/12)[33, 0, 54, 0, 36, 0, 18, 0, 3, 0, ...]^{T]}$.

8. $u_2 = u_4 = -5\alpha$, $v_3 = v_4 = -\alpha$, $\alpha = 10^{-4}/21$ gives exact stresses.

9. $\sigma_{xx} = \sigma_{yy} = 0$, $\sigma_{xy} = (21/8)10^5$.

10. Varying the number of elements gives for two pairs $\sigma_{xx} = -E\theta, 0, 0, E\theta$; for
     three pairs $\sigma_{xx} = -E\theta, -E\theta/3, -E\theta/3, E\theta/3, E\theta/3, E\theta$ and for four pairs:
     $\sigma_{xx} = -E\theta, -E\theta/2, -E\theta/2, 0, 0, E\theta/2, E\theta/2, E\theta$.

Index

Accuracy, 33
Adjacent elements, 68, 159, 165
Assembly process, 12–15, 55–6, 79–80
Axes, global and local, 4–12, 74–5, 125
  see also mapping

B matrix, 181
Banded matrix, 56–7
Bar, 1–3
Basis functions, 113, 116–19, 121, 142
Beam, 35, 109, 129
Bilinear functional, 103–7
Body forces, 192
Boundary conditions, 57, 69, 82, 101, 172

Calculus of variations, 26
Condensed force vector, 52
Condensed matrix, 52, 55, 73, 80–1, 124
Conductivity, 69, 110, 111
Conic curve, 151, 152
Continuity
  of mappings, 159–60, 164–5
  of solution, 30, 67–8
Convection boundary conditions, 60, 69, 82
Cross-section, 65, 125, 167
Curved element sides, 66, 151–3

Derivatives of the shape functions, 157, 181
Discretisation, 39, 65–7
Displacements, 6, 167–8
Distributed forces, 174
Divergence theorem, 99, 175

Elasticity, 22–4, 166–70
Electrostatics, 127
Element
  eight-noded, 148
  four-noded, 146
  isoparametric, 138
  linear, 30, 71, 142
  quadratic, 114, 145–9
  six-noded, 145
  subparametric, 138
  superparametric, 138
  three-noded, 144
Element force vector, 53, 85, 180
Element stiffness matrix, 6, 52, 85, 180
Energy
  gravitational, 25
  minimum, 25
  potential, 48
  strain, 48, 176
  total, 177
Energy principle, 48, 176
Equilibrium equations, 170
Error, 95
External forces, 35, 172

Finite differences, 184
Finite element equations, 14, 46, 77
Fluid flow, 126
Fourier series, 113
Functional, 26–9, 103–8, 176
Fundamental lemma, 96–8, 175

Gaussian quadrature, 137, 139–42
Gaussian elimination, 58
Global coordinates, 5
Global force vector, 47, 86
Global matrix, 46, 86
Gradient, 94
Gravitational energy, 25

Gravitational forces, 171, 192
Groundwater flow, 132

Heat flow, 59–61, 65–92
Heat source, 85
Hooke's law, 172
Howe truss, 1

Incomplete polynomials, 144
Infinite dimensional function space,
    112, 113
Integration by parts, 98–9, 175
Internal forces, 2, 14, 40
Internal stresses, 16, 169–70
Interpolation, 37, 142, 149
Inverse mapping, 154, 155
Irrotational fluid flow, 126
Isoparametric elements, 138

Jacobian, 136, 155
Jacobian matrix, 154–6
Joint, 2

Laplace's equation, 59, 65
Linear approximation, 30, 41, 61, 66
Linear functional, 103
Linear independence, 142
Linear mapping, 138, 156, 163
Linear triangle, 181
Local coordinates, 5

Mapping
    element, 134, 149–53, 159
    inverse, 154, 155
    linear, 142
Matrix
    banded, 56–7
    global, 47, 86
    Jacobian, 154–6
    stiffness, 6, 14, 52, 56, 70, 79, 181,
        186
Mesh generation, 160–2
Minimum energy principle, 26, 48

Natural boundary conditions, 57, 101
Newton–Cotes quadrature, 137
Non-linear mapping, 156
Normal strain, 167
Normal vector, 66, 99, 172
Numbering system, 2
Numerical integration, 137, 139–41

Optimisation process, 32

Piecewise interpolation, 31, 41, 68,
    115–19
Plane strain, 167
Plane stress, 167
Poisson's equation, 69, 125
Poisson's ratio, 172
Positive definite functional, 104, 105,
    176
Potential energy, 25, 48, 177
Prescribed displacement, 172
Prescribed temperature, 69
Pressure, 127
Principle of minimum energy, 26
Principle of virtual work, 176
Pyramid of polynomials, 143

Quadratic mapping, 138
Quadrature rules, 139–43
Quadrilateral element, 147–8

Reactions, 15, 38
Reduced set of equations, 188
Residual, 94–6, 174

Seepage, 127
Self-weight, 171
Shape functions, 41, 72, 113–15, 135,
    138, 142–9, 157
Shape functions derivatives, 157, 181
Shear modulus, 126
Shear strain, 169
Simpson's rule, 137
Singular matrix, 47
Six-noded element, 145

Spring, 50
Stable equilibrium, 48
Standard square, 141
Standard triangle, 142
Stationary point, 32
Stiffness matrix, 6, 14, 52, 56, 76, 79, 181, 186
Strain, 6, 167–9
Strain energy, 48, 174, 177
Stress, 6, 16, 167, 169–70
Subdomains, 39
Subparametric elements, 138
Superparametric elements, 138

Taylor theorem, 112
Temperature, 59, 65–92

Tension, 24, 34
Test functions, 98
Thermal conductivity, 59, 69
Three-noded element, 144
Torsion, 125
Total energy, 48, 177
Trace operator, 175
Trial functions, 95, 98, 119
Truss, 1

Variational form, 25, 69–70, 97
Variational principle, 22

Weak variational form, 100–102
Weighted residual methods, 95
Weights, 140, 143